THE CARTOON GUIDE TO
STATISTICS

Also by Larry Gonick

The Cartoon History of the Universe
The Cartoon Guide to Physics (with Art Huffman)
The Cartoon Guide to the Computer
The Cartoon Guide to Genetics (with Mark Wheelis)
The Cartoon History of the United States
The Cartoon Guide to (Non) Communication

THE CARTOON GUIDE TO STATISTICS

LARRY GONICK
& WOOLLCOTT SMITH

HarperResource
An Imprint of HarperCollins*Publishers*

HarperCollins books may be purchased for educational, business, or sales promotional use. For information please write: Special Markets Department, HarperCollins Publishers, Inc., 10 East 53rd Street, New York, NY 10022.

FIRST HARPERPERENNIAL EDITION

Illustrations by Larry Gonick

Library of Congress Cataloging-in-Publication Data

Gonick, Larry.
    The cartoon guide to statistics / Larry Gonick & Woollcott Smith.
    —1st HarperPerennial ed.
        p.   cm.
    Includes bibliographical references and index.
    ISBN 0-06-273102-5 (pbk.)
    1. Statistics—Caricatures and cartoons.   I. Smith, Woollcott, 1941–   .   II. Title.
QA276.12.G67   1993
519.5—dc20   92-54683

04 05 06 07 RRD 40 39 38 37 36 35 34 33 32

◆CONTENTS◆

Acknowledgments

WE WOULD LIKE TO THANK CAROL COHEN AT HARPERCOLLINS FOR SUGGESTING THIS PROJECT, OUR EDITOR ERICA SPABERG FOR PATIENTLY ENDURING THE LAST-MINUTE DASH TO THE DEADLINE, AND VICKY BIJUR, OUR LITERARY AGENT, FOR INITIATING THE GONICK/SMITH COLLABORATION BY INTRODUCING THE COAUTHORS.

WILLIAM FAIRLEY'S AND LEAH SMITH'S COMMENTS IMPROVED EARLIER DRAFTS OF THIS BOOK.

DONNA OKINO PROVIDED INVALUABLE ASSISTANCE AND ADVICE IN PRODUCING THE CARTOON PAGES. SHE SAYS THAT CREATING A CARTOON GUIDE IS HARDER THAN RUNNING A MARATHON, AND SHE SHOULD KNOW; SHE'S DONE BOTH.

THE ALTSYS CORPORATION CREATED FONTOGRAPHER, THE WONDERFUL SOFTWARE THAT ALLOWED US TO SIMULATE HAND-LETTERED TEXT AND FORMULAS ON THE MACINTOSH.

AND, SINCE EDUCATION IS ALWAYS A TWO-WAY STREET, A TIP OF THE HAT TO SMITH'S LONG-SUFFERING TEMPLE UNIVERSITY STUDENTS AND ESPECIALLY THE FALL '92 STUDY GROUP ORGANIZED BY ADRIANA TORRES. THE FUTURE IS THEIRS.

Chapter 1

WHAT IS STATISTICS?

WE MUDDLE THROUGH LIFE MAKING CHOICES
BASED ON *INCOMPLETE INFORMATION...*

SHOULD I HAVE THE SOUP? EVERYTHING ELSE IS SO *EXPENSIVE,* AND I DON'T KNOW WHO'S PAYING... ARE STATISTICIANS *STINGY?* I'VE NEVER GONE OUT WITH ONE BEFORE... THOUGH I ONCE KNEW A *VERY* GENEROUS ACCOUNTANT...

SHOULD I HAVE THE SOUP? *27* OUT OF THE *36* TIMES I'VE HAD IT, IT WAS PRETTY GOOD... BUT IS MONDAY THE REGULAR CHEF'S NIGHT OFF? AND WHAT IF ALL THE *AIR MOLECULES* IN THE ROOM SUDDENLY FLY UP TO THE CEILING?

MOST OF US LIVE COMFORTABLY WITH SOME LEVEL OF UNCERTAINTY.

WHAT MAKES STATISTICS UNIQUE IS ITS ABILITY TO *QUANTIFY* UNCERTAINTY, TO MAKE IT PRECISE. THIS ALLOWS STATISTICIANS TO MAKE *CATEGORICAL STATEMENTS*, WITH COMPLETE ASSURANCE—ABOUT THEIR LEVEL OF UNCERTAINTY!

THIS IS NOT JUST A MATTER OR ORDERING SOUP! STATISTICS ALSO INVOLVES MATTERS OF *LIFE AND DEATH*...

HEY—HAVE YOU EVER HAD THE SOUP HERE ON AN OFF NIGHT?

FOR EXAMPLE, IN 1986, THE SPACE SHUTTLE *CHALLENGER* EXPLODED, KILLING SEVEN ASTRONAUTS. THE DECISION TO LAUNCH IN 29-DEGREE WEATHER HAD BEEN MADE WITHOUT DOING A SIMPLE ANALYSIS OF PERFORMANCE DATA AT LOW TEMPERATURE.

OH...**THAT** PART OF THE CURVE !!!

A MORE POSITIVE EXAMPLE IS THE *SALK POLIO VACCINE*. IN 1954, VACCINE TRIALS WERE PERFORMED ON SOME 400,000 CHILDREN, WITH STRICT CONTROLS TO ELIMINATE BIASED RESULTS. GOOD STATISTICAL ANALYSIS OF THE RESULTS FIRMLY ESTABLISHED THE VACCINE'S EFFECTIVENESS, AND TODAY POLIO IS ALMOST UNKNOWN.

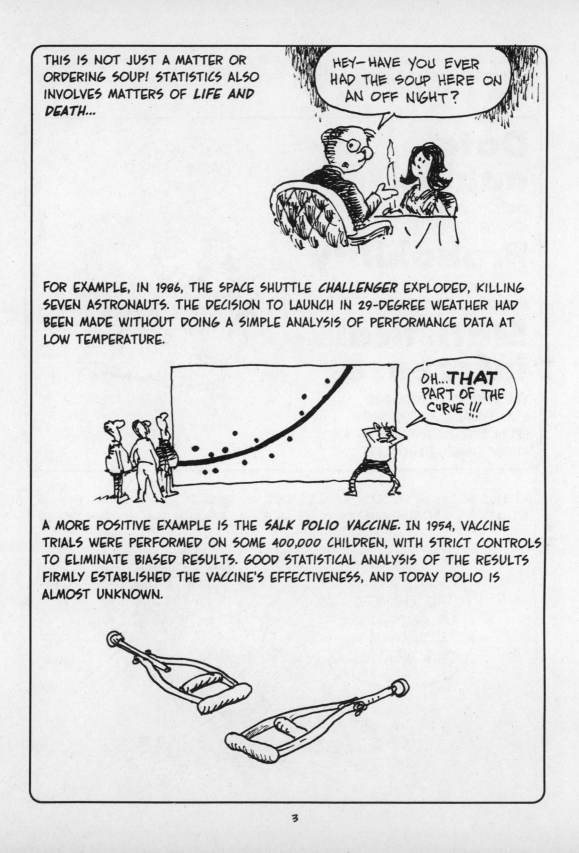

TO ACCOMPLISH THEIR FEATS OF MATHEMATICAL
LEGERDEMAIN, STATISTICIANS RELY ON THREE
RELATED DISCIPLINES:

Data analysis

THE GATHERING, DISPLAY, AND
SUMMARY OF DATA;

Probability

THE LAWS OF CHANCE, IN
AND OUT OF THE CASINO;

Statistical inference

THE SCIENCE OF DRAWING
STATISTICAL CONCLUSIONS
FROM SPECIFIC DATA, USING A
KNOWLEDGE OF PROBABILITY.

IN THIS BOOK, WE'LL LOOK AT ALL THREE, AS APPLIED TO A WIDE VARIETY OF
SITUATIONS WHERE STATISTICS PLAYS A CRUCIAL ROLE IN THE MODERN WORLD.

IN CHAPTER 2, WE'LL LOOK AT A SIMPLE DATA SET, THE REPORTED WEIGHTS OF A BUNCH OF COLLEGE STUDENTS.

IN CHAPTER 3, WE STUDY THE LAWS OF PROBABILITY IN THEIR BIRTHPLACE, THE GAMBLING DEN.

IN THE 17TH CENTURY ± 3% !!

CHAPTERS 4 AND 5 SHOW HOW TO DESCRIBE THE WORLD WITH *PROBABILITY MODELS*, USING THE CONCEPT OF THE *RANDOM VARIABLE*.

YOW! A SYMBOL!

CHAPTER 6 INTRODUCES ONE OF THE STATISTICIAN'S ESSENTIAL PROCEDURES, TAKING *SAMPLES* OF A LARGE POPULATION.

IN CHAPTER 7 AND BEYOND, WE DESCRIBE HOW TO MAKE STATISTICAL INFERENCES IN SUCH COMMON REAL-WORLD ARENAS AS *ELECTION POLLING, MANUFACTURING QUALITY CONTROL, MEDICAL TESTING, ENVIRONMENTAL MONITORING, RACIAL BIAS,* AND THE *LAW*.

IN SHORT, EVERYWHERE!

FINALLY, IN DISCUSSING STATISTICS, IT'S HARD TO AVOID MENTIONING ONE OTHER THING: THE WIDESPREAD *MISTRUST* OF STATISTICS IN THE WORLD TODAY. EVERYONE KNOWS ABOUT "LYING WITH STATISTICS," WHILE GOOD STATISTICAL ANALYSIS IS NEARLY IMPOSSIBLE TO FIND IN DAILY LIFE. WHAT'S ONE TO DO?

3 OUT OF 4 DOCTORS RECOMMEND NOT BELIEVING ANY STATEMENT BEGINNING .. WITH "3 OUT OF 4 DOCTORS...

OUR HUMBLE OPINION IS THAT *LEARNING A LITTLE MORE ABOUT THE SUBJECT* MIGHT NOT BE SUCH A BAD IDEA... AND THAT'S WHY WE WROTE THIS BOOK!

IN WHAT FOLLOWS, WE TRY TO PRESENT THE ELEMENTS OF STATISTICS AS GRAPHICALLY AND INTUITIVELY AS POSSIBLE. ALL YOU NEED TO GET THROUGH IT IS A LITTLE *PATIENCE*, SOME *THOUGHT*, AND A CERTAIN TOLERANCE FOR *ALGEBRA*—OR, IF NOT THAT, THEN MAYBE A *COURSE REQUIREMENT!!*

♦ CHAPTER 2 ♦
DATA DESCRIPTION

DATA ARE THE STATISTICIAN'S RAW MATERIAL, THE NUMBERS WE USE TO INTERPRET REALITY. ALL STATISTICAL PROBLEMS INVOLVE EITHER THE COLLECTION, DESCRIPTION, AND ANALYSIS OF DATA, OR *THINKING* ABOUT THE COLLECTION, DESCRIPTION, AND ANALYSIS OF DATA.

THIS CHAPTER CONCENTRATES ON DATA *DESCRIPTION*. HOW CAN WE REPRESENT DATA IN USEFUL WAYS? HOW CAN WE SEE UNDERLYING PATTERNS IN A HEAP OF NAKED NUMBERS? HOW CAN WE SUMMARIZE THE DATA'S BASIC SHAPE?

WELL, TO DESCRIBE DATA, THE FIRST THING YOU NEED IS SOME ACTUAL DATA TO DESCRIBE... SO LET'S COLLECT SOME DATA!

HERE IS SOME REAL DATA:
AS PART OF A CLASSROOM
EXPERIMENT, 92 PENN STATE
STUDENTS REPORTED THEIR
WEIGHT, WITH THESE
RESULTS:

MALES
140 145 160 190 155 165 150 190 195 138 160 155 153 145 170 175 175 170 180 135
170 157 130 185 190 155 170 155 215 150 145 155 155 150 155 150 180 160 135 160
130 155 150 148 155 150 140 180 190 145 150 164 140 142 136 123 155

FEMALES
140 120 130 138 121 125 116 145 150 112 125 130 120 130 131 120 118 125 135 125
118 122 115 102 115 150 110 116 108 95 125 133 110 150 108

GETTING RIGHT DOWN TO BUSINESS, WE DRAW A *DOT PLOT:* ONE DOT PER
STUDENT GOES OVER EACH STUDENT'S REPORTED WEIGHT.

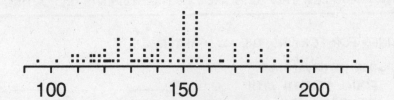
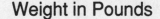

Weight in Pounds

YOU MAY SEE A *PROBLEM* HERE:
THE CLUMPS AT *150* AND *155*
POUNDS. THE STUDENTS TENDED
TO REPORT THEIR WEIGHT IN
FIVE-POUND INCREMENTS. IN
REAL-LIFE SITUATIONS LIKE THIS
ONE, SUCH ROUNDING OFF CAN
OBSCURE GENERAL PATTERNS IN
DATA... BUT FOR NOW, WE'LL JUST
WORK AROUND IT.

WE CAN SUMMARIZE THE DATA WITH A *FREQUENCY TABLE*. DIVIDE THE NUMBER LINE INTO INTERVALS AND COUNT THE NUMBER OF STUDENT WEIGHTS WITHIN EACH INTERVAL. THE *FREQUENCY* IS THE COUNT IN ANY GIVEN INTERVAL. THE *RELATIVE FREQUENCY* IS THE PROPORTION OF WEIGHTS IN EACH INTERVAL, I.E., IT'S THE FREQUENCY DIVIDED BY THE TOTAL NUMBER OF STUDENTS.

CLASS INTERVAL	MIDPOINT	FREQUENCY	RELATIVE FREQUENCY
87.5-102.4	95	2	.022
102.5-117.5	110	9	.098
117.5-132.4	125	19	.206
132.5-147.4	140	17	.185
147.5-162.4	155	27	.293
162.5-177.4	170	8	.087
177.5-192.4	185	8	.087
192.5-207.5	200	1	.011
207.5-222.4	215	1	.011
TOTAL		92	1.000

NOTE: WE KEPT THE INTERVAL BOUNDARIES AWAY FROM THOSE TROUBLESOME 5-POUND MULTIPLES. THIS GETS AROUND THE STUDENTS' REPORTING BIAS.

GUIDELINES FOR FORMING THE CLASS INTERVALS:

1) USE INTERVALS OF EQUAL LENGTH WITH MIDPOINTS AT CONVENIENT ROUND NUMBERS.

2) FOR A SMALL DATA SET, USE A SMALL NUMBER OF INTERVALS.

3) FOR A LARGE DATA SET, USE MORE INTERVALS!

10

IN THE FREQUENCY TABLE, WE ARE SHOWING HOW MANY DATA POINTS ARE "AROUND" EACH VALUE. WE CAN GRAPH THIS INFORMATION, TOO. THE RESULTING BAR GRAPH IS CALLED A *HISTOGRAM*. EACH BAR COVERS AN INTERVAL AND IS CENTERED AT THE MIDPOINT. THE BAR'S HEIGHT IS THE NUMBER OF DATA POINTS IN THE INTERVAL.

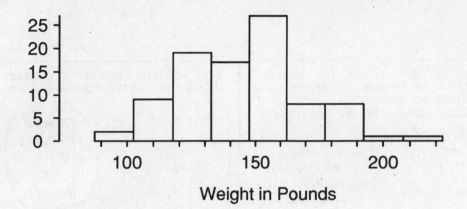

WE CAN ALSO DRAW A *RELATIVE FREQUENCY HISTOGRAM*, PLOTTING THE RELATIVE FREQUENCY AGAINST THE WEIGHT. IT LOOKS EXACTLY THE SAME, EXCEPT FOR THE VERTICAL SCALE.

11

THE STATISTICIAN *JOHN TUKEY* INVENTED A QUICK WAY TO SUMMARIZE DATA AND STILL KEEP THE INDIVIDUAL DATA POINTS. IT'S CALLED THE **STEM-AND-LEAF** DIAGRAM.

FOR THE WEIGHT DATA, THE STEM IS A COLUMN OF NUMBERS, CONSISTING OF THE WEIGHT DATA COUNTED BY TENS (I.E., WE LEAVE OFF THE LAST DIGIT).

9
10
11
12
13
14
15
16
17
18
19
20
21

I.E., 90 POUNDS, 100 POUNDS, ETC.

NOW ADD THE FINAL DIGIT OF EACH WEIGHT IN THE APPROPRIATE ROW:

STEM : LEAVES
9 :
10 :
11 : 628
12 : 0155005
13 : 080015
14 : 05
15 : 0
16 :
17 :
18 :
19 :
20 :
21 :

MEANING THERE ARE WEIGHTS OF 116, 112, 118, 120, ETC.

FILLED IN, IT LOOKS LIKE THIS:

9 : 5
10 : 288
11 : 628855060
12 : 01553005525
13 : 8500850600153
14 : 05505580502
15 : 50537055055050505000500
16 : 050004
17 : 055000
18 : 0500
19 : 00500
20 :
21 : 5

AND FINALLY, PUT THE "LEAVES" IN ORDER.

9 : 5
10 : 288
11 : 002556688
12 : 00012355555
13 : 0000013555688
14 : 00002555558
15 : 000000000035555555555557
16 : 000045
17 : 000055
18 : 0005
19 : 00005
20 :
21 : 5

ALL THOSE ZEROES AND FIVES CLEARLY SHOW THE STUDENTS' REPORTING BIAS!

GOOD GRAPHIC DISPLAY IS PART ART AND PART SCIENCE

AND SOMETIMES, PART POLITICS!

CRUSADING NURSE *FLORENCE NIGHTINGALE* COMPILED *MORTALITY STATISTICS* FROM BRITISH MILITARY HOSPITALS, PRODUCING SHOCKING HISTOGRAMS LIKE THIS ONE: THE RADIAL AXIS INDICATES DEATHS—IN HOSPITALS AS WELL AS ON THE BATTLEFIELD— OF BRITISH SOLDIERS IN THE CRIMEAN WAR.

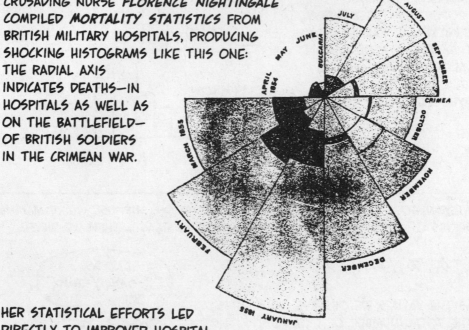

HER STATISTICAL EFFORTS LED DIRECTLY TO IMPROVED HOSPITAL CONDITIONS AND A REDUCTION IN THE DEATH RATE.

SAVED BY STATISTICS!

SUMMARY STATISTICS

NOW WE MOVE FROM PICTURES TO FORMULAS. OUR OBJECT IS TO GET SOME SIMPLE MEASUREMENTS OF THE CRUDEST CHARACTERISTICS OF A SET OF DATA....

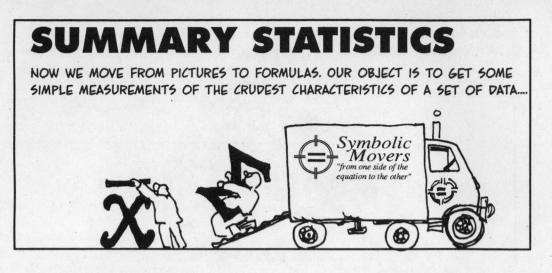

ANY SET OF MEASUREMENTS HAS TWO IMPORTANT PROPERTIES: THE *CENTRAL OR TYPICAL VALUE*, AND THE *SPREAD* ABOUT THAT VALUE. YOU CAN SEE THE IDEA IN THESE HYPOTHETICAL HISTOGRAMS.

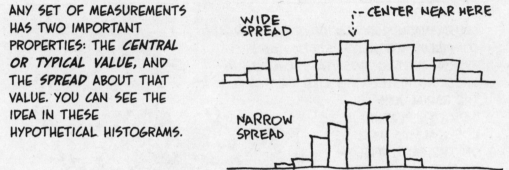

WE CAN GO A LONG WAY WITH A LITTLE NOTATION. SUPPOSE WE'RE MAKING A SERIES OF OBSERVATIONS... n OF THEM, TO BE EXACT... THEN WE WRITE

$$x_1, x_2, x_3, \dots x_n$$

AS THE VALUES WE OBSERVE. THUS, n IS THE TOTAL NUMBER OF DATA POINTS, AND x_4 (SAY) IS THE VALUE OF THE FOURTH DATA POINT.

READ AS "X-ONE, X-TWO," ETC.

AN *ARRAY* IS A TABLE OF DATA:

OBSERVATION	1	2	3	4	....	n
DATA VALUE	x_1	x_2	x_3	x_4	....	x_n

A SMALL SET OF $n = 5$ DATA POINTS MAKES THE BOOKKEEPING EASY. SUPPOSE, FOR EXAMPLE, WE ASK FIVE PEOPLE HOW MANY HOURS OF *TELEVISION* THEY WATCH IN A WEEK... AND GET THE FOLLOWING ARRAY:

OBSERVATION	1	2	3	4	5
DATA VALUE	5	7	3	38	7

THEN $x_1 = 5$, $x_2 = 7$, $x_3 = 3$, $x_4 = 38$, AND $x_5 = 7$.

WHAT'S THE "CENTER" OF THESE DATA? THERE ARE ACTUALLY SEVERAL DIFFERENT WAYS TO MEASURE IT. WE'LL LOOK AT JUST TWO OF THEM.

THE # MEAN (OR "AVERAGE")

THE *MEAN* OR AVERAGE VALUE IS REPRESENTED BY $\bar{x}$... IT'S OBTAINED BY ADDING ALL THE DATA AND DIVIDING BY THE NUMBER OF OBSERVATIONS:

$$\bar{x} = \frac{\text{SUM OF DATA}}{n}$$

$$= \frac{x_1 + x_2 + ... + x_n}{n}$$

FOR OUR EXAMPLE,

$$\bar{x} = \frac{5 + 7 + 3 + 38 + 7}{5} = \frac{60}{5}$$

$$= \mathbf{12} \text{ HOURS}$$

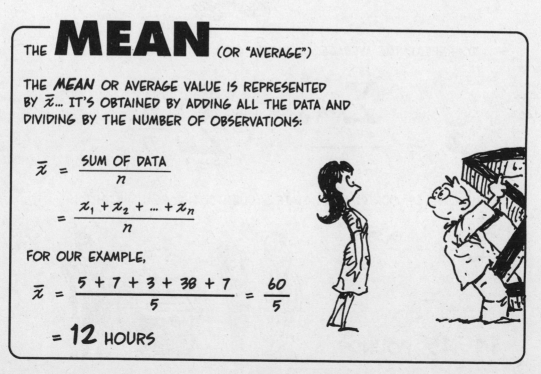

WE HAVE A SHORTHAND FOR THAT $x_1 + x_2 + ... + x_n$ USING THE GREEK CAPITAL LETTER *SIGMA*, FOR SUMMATION:

$$\Sigma$$

FOR THE SUM $x_1 + x_2 + ... + x_n$ WE WRITE

$$\sum_{i=1}^{n} x_i$$

AND READ IT AS "THE SUM OF x_i AS i GOES FROM 1 TO n."

SAY IT TEN TIMES AND YOU'LL NEVER FORGET IT...

ALL RIGHT! NOW WE LOOKIN' LIKE A *STATISTICS* BOOK!

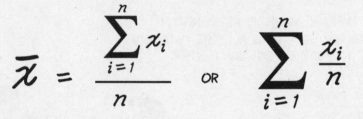

SO... TO REPEAT, THE AVERAGE, OR *MEAN*, OF A SET OF DATA x_i IS

$$\bar{x} = \frac{\sum_{i=1}^{n} x_i}{n} \quad \text{OR} \quad \sum_{i=1}^{n} \frac{x_i}{n}$$

IN THE CASE OF OUR 92 PENN STATE STUDENTS, THE MEAN WEIGHT IS

$$\sum_{i=1}^{92} \frac{x_i}{92} = \frac{13,354}{92} = $$

145.15 POUNDS

THE **MEDIAN**  IS ANOTHER KIND OF CENTER: THE "MIDPOINT" OF THE DATA, LIKE THE "MEDIAN STRIP" IN A ROAD.

TO FIND THE MEDIAN VALUE OF A DATA SET, WE ARRANGE THE DATA IN ORDER FROM SMALLEST TO LARGEST. THE MEDIAN IS THE VALUE IN THE MIDDLE.

3   5   7   7   38

↑

THE MEDIAN

IF THE NUMBER OF POINTS IS *EVEN*—IN WHICH CASE THERE IS NO MIDDLE, WE AVERAGE THE TWO VALUES AROUND THE MIDDLE... SO IF THE DATA ARE

3   5   7   7

↑

MIDDLE SPACE

WE AVERAGE 5 AND 7 TO GET

$$\frac{5 + 7}{2} = 6$$

THIS GIVES US A GENERAL RULE: ORDER THE DATA FROM SMALLEST TO LARGEST.

IF THE NUMBER OF DATA POINTS IS *ODD*, THE MEDIAN IS THE MIDDLE DATA POINT.

IF THE NUMBER OF POINTS IS *EVEN*, THE MEDIAN IS THE AVERAGE OF THE TWO DATA POINTS NEAREST THE MIDDLE.

JUST AS THE MEDIAN STRIP'S POSITION IS THERE, BUT NOT THE STRIP...

FOR THE $n=92$ STUDENT WEIGHTS, WE CAN FIND THE MEDIAN FROM THE ORDERED STEM-AND-LEAF DIAGRAM: JUST COUNT TO THE 46TH OBSERVATION. THE MEDIAN IS

$$\frac{x_{46} + x_{47}}{2} = \frac{145 + 145}{2}$$

$$= \mathbf{145} \text{ POUNDS}$$

```
 9 : 5
10 : 288
11 : 002556688
12 : 00012355555
13 : 0000013555688
14 : 00002555 55 8
15 : 00000000003555555555557
16 : 000045
17 : 000055
18 : 0005
19 : 00005
20 :
21 : 5
```

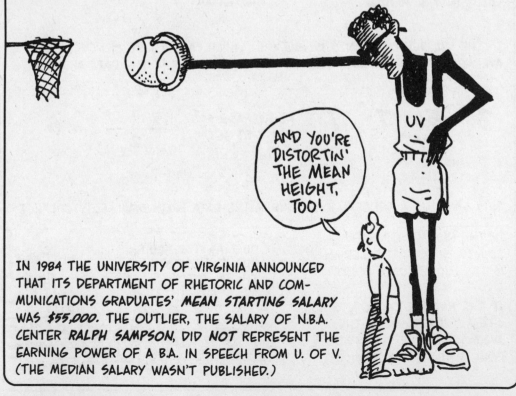

WHY MORE THAN ONE MEASURE OF THE CENTER? EACH HAS ADVANTAGES. FOR EXAMPLE, THE *MEDIAN* IS NOT SENSITIVE TO *OUTLIERS*, OR EXTREME VALUES NOT TYPICAL OF THE REST OF THE DATA. SUPPOSE IN OUR SMALL TV-WATCHING GROUP, ONE PERSON WATCHES *200* HOURS PER WEEK. THEN OUR DATA ARE 3, 5, 7, 7, 200. THE MEDIAN, 7, IS UNCHANGED, BUT THE MEAN IS NOW $\bar{x} = \mathbf{45.8}!$

AND YOU'RE DISTORTIN' THE MEAN HEIGHT, TOO!

IN 1984 THE UNIVERSITY OF VIRGINIA ANNOUNCED THAT ITS DEPARTMENT OF RHETORIC AND COMMUNICATIONS GRADUATES' *MEAN STARTING SALARY* WAS *$55,000*. THE OUTLIER, THE SALARY OF N.B.A. CENTER *RALPH SAMPSON*, DID *NOT* REPRESENT THE EARNING POWER OF A B.A. IN SPEECH FROM U. OF V. (THE MEDIAN SALARY WASN'T PUBLISHED.)

MEASURES OF
S P R E A D

BESIDES KNOWING THE
CENTRAL POINT OF A DATA
SET, WE'D ALSO LIKE TO
DESCRIBE THE DATA'S
SPREAD, OR HOW FAR
FROM THE CENTER THE
DATA TEND TO RANGE.
FOR INSTANCE, IF THE
STUDENTS ALL WEIGHED
EXACTLY 145 POUNDS,
THERE WOULD BE NO
SPREAD AT ALL.
NUMERICALLY, THE SPREAD
WOULD BE *ZERO*, AND THE
HISTOGRAM WOULD BE
SKINNY.

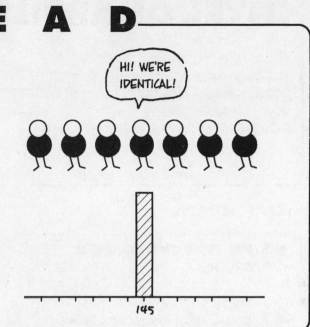

HI! WE'RE
IDENTICAL!

145

BUT IF MANY OF THE STUDENTS WERE VERY LIGHT AND/OR VERY HEAVY,
OBVIOUSLY WE'D SEE SOME SPREAD—SAY, IF THE *FOOTBALL TEAM* WAS PART
OF THE SAMPLE...

THE HISTOGRAM WOULD BE WIDER, SOMETHING LIKE THIS:

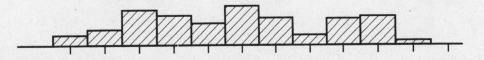

AGAIN, THERE'S MORE THAN ONE WAY TO MEASURE A SPREAD. ONE WAY IS

INTERQUARTILE RANGE

THE IDEA IS TO DIVIDE THE DATA INTO FOUR EQUAL GROUPS AND SEE HOW FAR APART THE EXTREME GROUPS ARE.

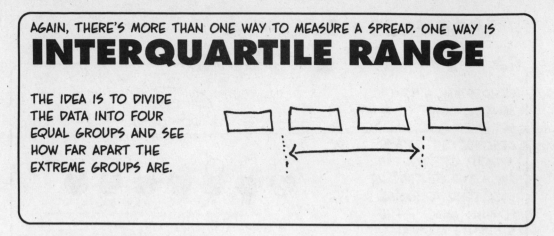

HERE'S THE RECIPE:

1) PUT THE DATA IN NUMERICAL ORDER.

2) DIVIDE THE DATA INTO TWO EQUAL HIGH AND LOW GROUPS AT THE MEDIAN. (IF THE MEDIAN IS A DATA POINT, INCLUDE IT IN BOTH THE HIGH AND LOW GROUPS.)

3) FIND THE MEDIAN OF THE LOW GROUP. THIS IS CALLED THE FIRST QUARTILE, OR Q_1.

4) THE MEDIAN OF THE HIGH GROUP IS THE THIRD QUARTILE, OR Q_3.

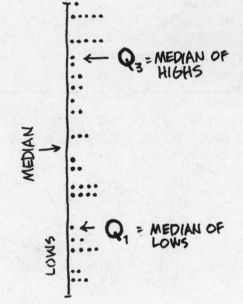

Q_3 = MEDIAN OF HIGHS

MEDIAN

Q_1 = MEDIAN OF LOWS

LOWS

NOW THE INTERQUARTILE RANGE (IQR) IS THE DISTANCE (OR DIFFERENCE) BETWEEN THEM:

$$IQR = Q_3 - Q_1$$

HERE'S THE WEIGHT DATA WITH THE MIDPOINTS OF THE HIGH AND LOW GROUPS EMPHASIZED:

```
 9 : 5
10 : 288
11 : 002556688   ←
12 : 00012355555
13 : 0000013555688
14 : 00002555558
15 : 0000000000035555555555557
16 : 000045
17 : 000055              ↑
18 : 0005
19 : 00005
20 :
21 : 5
```

AND WE SEE THAT

$$\text{IQR} = 156 - 125$$
$$= 31 \text{ POUNDS}$$

AGAIN, THIS IS THE DIFFERENCE BETWEEN THE MEDIAN HEAVY STUDENT AND MEDIAN LIGHT ONE.

JOHN TUKEY INVENTED ANOTHER KIND OF DISPLAY TO SHOW OFF THE IQR, CALLED A *BOX AND WHISKERS* PLOT. THE BOX'S ENDS ARE THE QUARTILES Q_1 AND Q_3. WE DRAW THE MEDIAN INSIDE THE BOX.

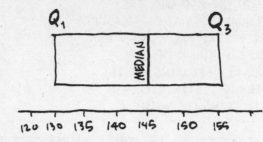

IF A POINT IS MORE THAN 1.5 IQR FROM AN END OF THE BOX, IT'S AN *OUTLIER*. DRAW THE OUTLIERS INDIVIDUALLY.

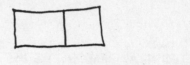

FINALLY, EXTEND "WHISKERS" OUT TO THE FARTHEST POINTS THAT ARE NOT OUTLIERS (I.E., WITHIN 1.5 IQR OF THE QUARTILES).

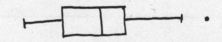
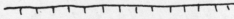

BOX-AND-WHISKERS PLOTS ARE ESPECIALLY GOOD FOR SHOWING OFF DIFFERENCES BETWEEN GROUPS.

THE STANDARD MEASURE OF SPREAD IS THE

STANDARD DEVIATION

UNLIKE THE IQR, WHICH IS BASED ON MEDIANS, THE STANDARD DEVIATION MEASURES THE SPREAD FROM THE *MEAN.* YOU CAN THINK OF IT, ROUGHLY SPEAKING, AS THE AVERAGE DISTANCE OF THE DATA FROM THE MEAN $\bar{x}$...

EXCEPT THAT WE USE THE *SQUARES* OF THE DISTANCES INSTEAD. THAT IS, IF THE SQUARED DISTANCE OF POINT x_i TO $\bar{x}$ IS $(x_i - \bar{x})^2$, THEN

$$\text{AVERAGE SQUARED DISTANCE} = \frac{1}{n}\sum_{i=1}^{n}(x_i - \bar{x})^2$$

FOR TECHNICAL REASONS, WE USE $n-1$ IN THE DENOMINATOR RATHER THAN n, AND DEFINE THE *SAMPLE VARIANCE* S^2 AS

$$S^2 = \frac{1}{n-1}\sum_{i=1}^{n}(x_i - \bar{x})^2$$

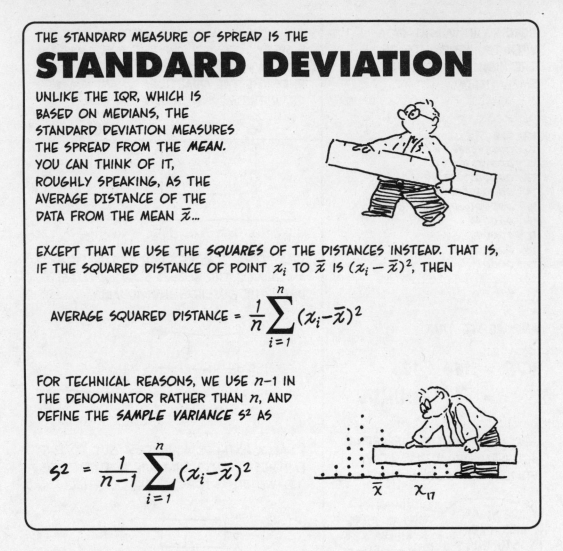

FOR THE DATA SET $\{3\ \ 5\ \ 7\ \ 7\ \ 38\}$, WITH $\bar{x} = 12$ AND $n = 5$ WE CALCULATE THE VARIANCE:

$$S^2 = \frac{(3-12)^2 + (5-12)^2 + (7-12)^2 + (7-12)^2 + (38-12)^2}{(5-1)}$$

$$= \frac{81 + 49 + 25 + 25 + 676}{4}$$

$$= 214$$

THE LARGE VARIANCE HERE REFLECTS THE WIDE SPREAD IN THE DATA...

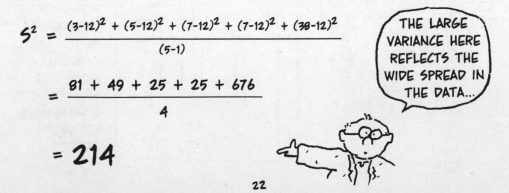

BUT A SPREAD MEASURE SHOULD HAVE THE SAME *UNITS* AS THE ORIGINAL DATA. IN THE EXAMPLE OF WEIGHTS, THE VARIANCE S^2 IS MEASURED IN POUNDS *SQUARED*... OOOPS!

THE OBVIOUS THING TO DO IS TO TAKE THE SQUARE ROOT, AND SO WE DO... TO DEFINE:

STANDARD DEVIATION

$$S = \sqrt{S^2} = \sqrt{\frac{1}{n-1}\sum_{i=1}^{n}(x_i - \bar{x})^2}$$

WHICH, FOR OUR SIMPLE DATA SET, IS

$$S = \sqrt{214} = 14.63$$

NNGH!

WHO THE ✖⊘♯ CAN REMEMBER HOW TO TAKE SQUARE ROOTS ?!!

EVEN FOR SMALL DATA SETS, THE ARITHMETIC CAN BE TEDIOUS! SO NOWADAYS, WE JUST HIT THE S BUTTON ON THE HAND CALCULATOR, OR CONSULT THE DATA REPORT GENERATED BY A COMPUTER SOFTWARE PACKAGE.

Properties of $\overline{X}$ and S

THE MEAN AND STANDARD
DEVIATION ARE VERY GOOD
FOR SUMMARIZING THE
PROPERTIES OF FAIRLY
*SYMMETRICAL HISTOGRAMS
WITHOUT OUTLIERS*—I.E.,
HISTOGRAMS SHAPED LIKE
MOUNDS.

$\overline{x}-s \qquad \overline{x} \qquad \overline{x}+s$

A SHAPE TO REMEMBER!

IT'S OFTEN USEFUL TO KNOW HOW MANY STANDARD DEVIATIONS A DATA POINT IS FROM THE MEAN. WE DEFINE *z-SCORES*, OR STANDARDIZED SCORES, AS DISTANCE FROM $\overline{x}$ *PER STANDARD DEVIATION*.

$$z_i = \frac{x_i - \overline{x}}{s} \quad \text{FOR EACH } i.$$

A z-SCORE OF +2 MEANS THAT AN OBSERVATION IS *TWO STANDARD DEVIATIONS ABOVE THE MEAN*. FOR THE WEIGHT DATA ($\overline{x} = 145.2$ AND $S = 23.7$), WE CAN PLOT THE DATA ON THE ORIGINAL x-AXIS IN POUNDS AND THE z-SCORE AXIS SIMULTANEOUSLY.

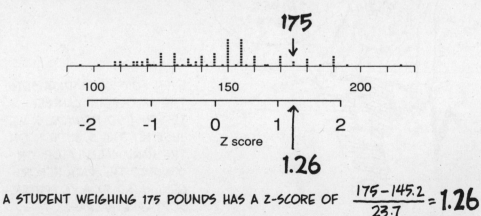

A STUDENT WEIGHING 175 POUNDS HAS A z-SCORE OF $\dfrac{175 - 145.2}{23.7} = 1.26$

24

an EMPIRICAL RULE:

FOR NEARLY SYMMETRIC MOUND-SHAPED DATA SETS, APPROXIMATELY *68%* OF THE DATA IS WITHIN *ONE* STANDARD DEVIATION OF THE MEAN AND *95%* OF THE DATA IS WITHIN *TWO* STANDARD DEVIATIONS OF THE MEAN.

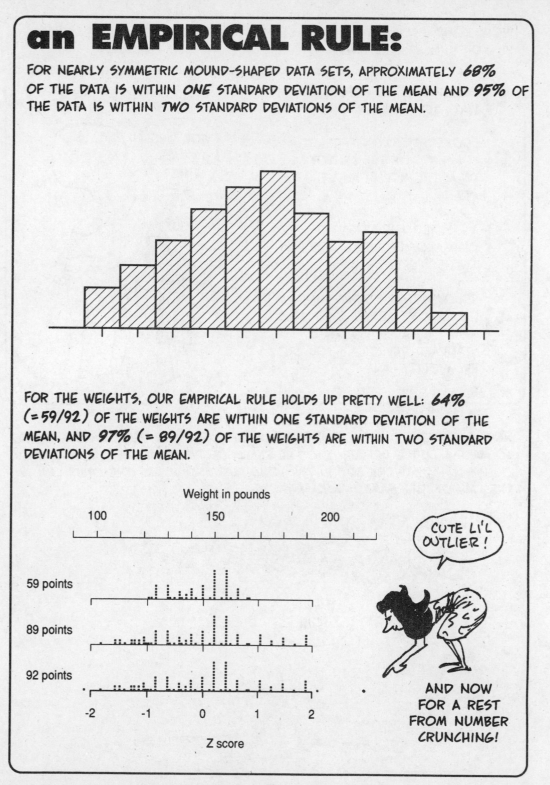

FOR THE WEIGHTS, OUR EMPIRICAL RULE HOLDS UP PRETTY WELL: *64%* (=59/92) OF THE WEIGHTS ARE WITHIN ONE STANDARD DEVIATION OF THE MEAN, AND *97%* (= 89/92) OF THE WEIGHTS ARE WITHIN TWO STANDARD DEVIATIONS OF THE MEAN.

Weight in pounds

100   150   200

59 points

89 points

92 points

-2   -1   0   1   2

Z score

CUTE LI'L OUTLIER!

AND NOW FOR A REST FROM NUMBER CRUNCHING!

WE'VE COME A LONG WAY IN THIS CHAPTER! STARTING WITH A UNORGANIZED PILE OF NUMBERS, WE HAVE:

1) FOUND SEVERAL DIFFERENT WAYS TO DISPLAY THEM

2) LOOKED AT TWO DIFFERENT CONCEPTS OF THE CENTER OF DATA, THE MEDIAN AND THE MEAN

3) MEASURED THE SPREAD OF THE DATA AROUND THE CENTER IN TWO DIFFERENT WAYS

4) ENCOUNTERED MOUND-SHAPED HISTOGRAMS AND Z, A VARIABLE THAT INDICATES HOW MANY STANDARD DEVIATIONS YOU ARE FROM THE MEAN.

WOW! WE DID ALL THAT?

I'M 98% SURE OF IT...

NOW, IN ORDER TO PROBE THE BEHAVIOR OF DATA MORE DEEPLY, WE'RE GOING TO MAKE A LITTLE DETOUR INTO THE REALM OF *RANDOMNESS*... A LAND WHERE THINGS *ALWAYS* WORK OUT IN THE LONG RUN, AND WHERE THE ONLY LAW IS THE LAW OF THE *GAMBLING CASINO*...

ROLL 'EM!

♦ Chapter 3 ♦
PROBABILITY

N OTHING IN LIFE IS CERTAIN. IN EVERYTHING WE DO, WE GAUGE THE CHANCES OF SUCCESSFUL OUTCOMES, FROM BUSINESS TO MEDICINE TO THE WEATHER. BUT FOR MOST OF HUMAN HISTORY, *PROBABILITY*, THE FORMAL STUDY OF THE LAWS OF CHANCE, WAS USED FOR ONLY ONE THING: *GAMBLING.*

NOBODY KNOWS WHEN GAMBLING BEGAN. IT GOES BACK AT LEAST AS FAR AS **ANCIENT EGYPT,** WHERE SPORTING MEN AND WOMEN USED FOUR-SIDED "ASTRAGALI" MADE FROM ANIMAL HEELBONES.

BURY ME WITH MY ASTRAGALI... I WANT TO CHEAT DEATH!

THE ROMAN EMPEROR **CLAUDIUS** (10 BCE-54 CE) WROTE THE FIRST KNOWN TREATISE ON GAMBLING. UNFORTUNATELY, THIS BOOK, "HOW TO WIN AT DICE," WAS LOST.

RULE I: LET CAESAR WIN IV OUT OF V!

MODERN DICE GREW POPULAR IN THE MIDDLE AGES, IN TIME FOR A RENAISSANCE RAKE, THE **CHEVALIER DE MERE,** TO POSE A MATHEMATICAL PUZZLER:

WHAT'S LIKELIER: ROLLING AT LEAST **ONE SIX** IN **FOUR THROWS OF A SINGLE DIE,** OR ROLLING AT LEAST **ONE DOUBLE SIX** IN **24 THROWS OF A PAIR OF DICE?**

THE CHEVALIER REASONED
THAT THE AVERAGE NUMBER
OF SUCCESSFUL ROLLS WAS
THE SAME FOR BOTH GAMBLES:

CHANCE OF ONE SIX $= \frac{1}{6}$

AVERAGE NUMBER IN
FOUR ROLLS $= 4 \cdot \left(\frac{1}{6}\right) = \frac{2}{3}$

CHANCE OF DOUBLE
SIX IN ONE ROLL $= \frac{1}{36}$

AVERAGE NUMBER IN
24 ROLLS $= 24 \cdot \left(\frac{1}{36}\right) = \frac{2}{3}$

WHY, THEN, DID HE *LOSE MORE OFTEN* WITH THE SECOND GAMBLE???

DE MERE PUT THE QUESTION TO HIS FRIEND, THE GENIUS *BLAISE PASCAL* (1623-1666).

ALTHOUGH PASCAL HAD EARLIER GIVEN UP MATHEMATICS AS A FORM OF SEXUAL INDULGENCE (!), HE AGREED TO TACKLE DE MERE'S PROBLEM.

PASCAL WROTE HIS FELLOW GENIUS *PIERRE DE FERMAT*, AND WITHIN A FEW LETTERS, THE TWO HAD WORKED OUT THE THEORY OF PROBABILITY IN ITS MODERN FORM—EXCEPT, OF COURSE, FOR THE CARTOONS.

BASIC DEFINITIONS

AS OUR GAMBLER PLAYS A GAME, WE PLAY SCIENTIST, OBSERVING THE OUTCOME:

A **random experiment** IS THE PROCESS OF OBSERVING THE OUTCOME OF A CHANCE EVENT.

THE **elementary outcomes** ARE ALL POSSIBLE RESULTS OF THE RANDOM EXPERIMENT.

THE **sample space** IS THE SET OR COLLECTION OF ALL THE ELEMENTARY OUTCOMES.

WHAT GAME? DICE? CHEMIN·DE·FER?

LET'S FLIP A COIN!

IF THE EVENT WAS A COIN TOSS, FOR EXAMPLE, THE *RANDOM EXPERIMENT* CONSISTS OF RECORDING ITS OUTCOME...

THE *ELEMENTARY OUTCOMES* ARE HEADS AND TAILS...

H   T

AND THE *SAMPLE SPACE* IS THE SET WRITTEN

$$\{H, T\}$$

AND IF DICE IS YOUR GAME?

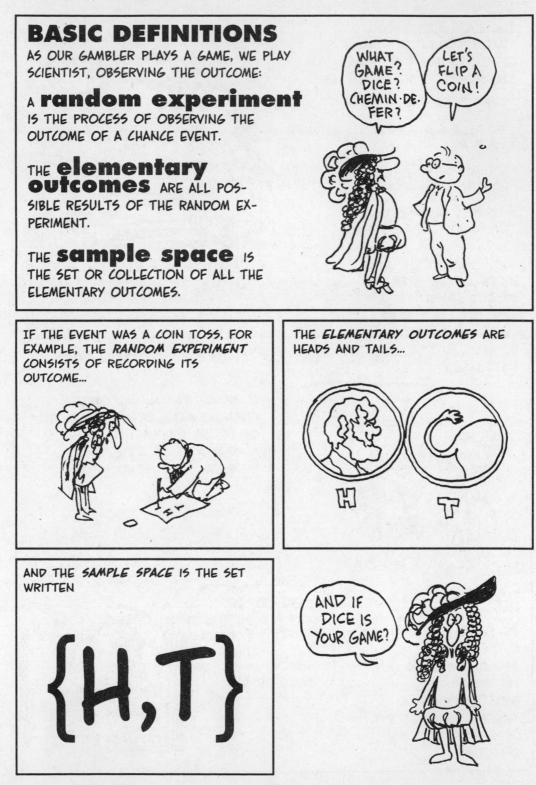

THE SAMPLE SPACE OF THE THROW OF A *SINGLE DIE* IS A LITTLE BIGGER.

AND FOR A *PAIR* OF DICE, THE SAMPLE SPACE LOOKS LIKE THIS (WE MAKE ONE DIE WHITE AND ONE BLACK TO TELL THEM APART):

THIS SAMPLE SPACE HAS *36* (6X6) ELEMENTARY OUTCOMES. FOR THREE DICE, THE SPACE WOULD HAVE 216 ENTRIES, AS IN THIS 6X6X6 STACK. AND *FOUR DICE?*

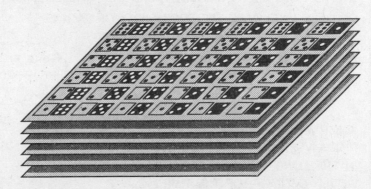

ENOUGH!

AT SOME POINT, WE HAVE TO STOP LISTING, AND START THINKING...

NOW LET'S IMAGINE A RANDOM EXPERIMENT WITH n ELEMENTARY OUTCOMES $O_1, O_2, \ldots O_n$. WE WANT TO ASSIGN A **NUMERICAL WEIGHT**, OR **PROBABILITY**, TO EACH OUTCOME, WHICH MEASURES THE LIKELIHOOD OF ITS OCCURRING. WE WRITE THE PROBABILITY OF O_i AS $P(O_i)$.

FOR EXAMPLE, IN A FAIR COIN TOSS, HEADS AND TAILS ARE EQUALLY LIKELY, AND WE ASSIGN THEM BOTH THE PROBABILITY .5.

$$P(H) = P(T) = .5$$

EACH OUTCOME COMES UP HALF THE TIME. ASK ANY **FOOTBALL PLAYER!**

IN THE ROLL OF **TWO DICE**, THERE ARE **36** ELEMENTARY OUTCOMES, ALL EQUALLY LIKELY, SO THE PROBABILITY OF EACH IS $\frac{1}{36}$.

FOR INSTANCE,

$$P(\text{BLACK } 5, \text{ WHITE } 2) = \frac{1}{36}$$

WHICH MEANS: IF YOU ROLLED THE DICE A VERY LARGE NUMBER OF TIMES, IN THE LONG RUN THIS OUTCOME WOULD OCCUR $\frac{1}{36}$ OF THE TIME.

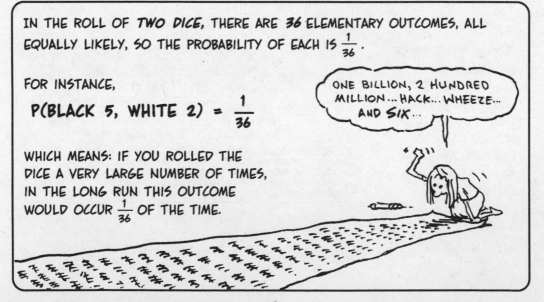

WHAT IF OUR GAMBLER *CHEATS* AND THROWS A *LOADED DIE?* FOR THE SAKE OF ARGUMENT, SUPPOSE THAT NOW A ONE COMES UP **25%** OF THE TIME (IN THE LONG RUN).

THE SAMPLE SPACE IS THE SAME AS FOR A FAIR DIE

$$\{1, 2, 3, 4, 5, 6\}$$

BUT THE PROBABILITIES ARE DIFFERENT. NOW $P(1) = .25$ AND THE REMAINING PROBABILTIES ADD UP TO .75. IF 2, 3, 4, 5, AND 6 WERE ALL EQUALLY LIKELY, THEN EACH ONE WOULD HAVE PROBABILITY $.15 = \frac{1}{5}(.75)$

.25   .15   .15   .15   .15   .15

I CAN WORK WITH THAT!

IN GENERAL, ELEMENTARY OUTCOMES NEED NOT HAVE EQUAL PROBABILITY.

THE PROBABILITY OF PRECIPITATION IS 20%...

THE PROBABILITY OF A WALK IS 5%...

NOW WHAT CAN WE SAY ABOUT THE PROBABILITIES $P(O_i)$ IN AN ARBITRARY RANDOM EXPERIMENT? FIRST OF ALL,

$$P(O_i) \geqslant 0$$

PROBABILITIES ARE *NEVER NEGATIVE*. A PROBABILITY OF ZERO MEANS AN EVENT CAN'T HAPPEN. LESS THAN ZERO WOULD BE MEANINGLESS.

SECOND, IF AN EVENT IS *CERTAIN* TO HAPPEN, WE ASSIGN IT PROBABILITY 1. (IN THE LONG RUN, THAT'S THE PROPORTION OF TIMES IT WILL *OCCUR*!)

 IN PARTICULAR, THE *TOTAL PROBABILITY OF THE SAMPLE* SPACE MUST BE 1. IF WE DO THE EXPERIMENT, *SOMETHING* IS BOUND TO HAPPEN!

PUT THESE TWO TOGETHER, AND YOU HAVE THE *CHARACTERISTIC PROPERTIES OF PROBABILITY*:

$$P(O_i) \geqslant 0$$
$$P(O_1) + P(O_2) + \ldots + P(O_n) = 1$$

PROBABILITY IS *NON-NEGATIVE*

TOTAL PROBABILITY OF ALL ELEMENTARY OUTCOMES IS *ONE.*

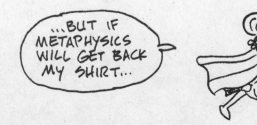

LIKE A CLEVER POLITICIAN, WE HAVE AVOIDED CERTAIN *UNPLEASANT QUESTIONS*, SUCH AS A) WHAT DOES PROBABILITY *MEAN?* AND B) *HOW* DO WE ASSIGN PROBABILITIES TO OUTCOMES?

B-DUH, B-DUH... LET'S DISCUSS SOMETHING EASIER, LIKE GAYS IN THE MILITARY...

HERE ARE SOME APPROACHES THAT HAVE BEEN TAKEN:

Classical PROBABILITY:
BASED ON GAMBLING IDEAS, THE FUNDAMENTAL ASSUMPTION IS THAT THE GAME IS FAIR AND ALL ELEMENTARY OUTCOMES HAVE THE SAME PROBABILITY.

C'MON! DADDY NEEDS A NEW THEORY!

Relative Frequency:
WHEN AN EXPERIMENT *CAN* BE REPEATED, THEN AN EVENT'S PROBABILITY IS THE PROPORTION OF TIMES THE EVENT OCCURS IN THE LONG RUN.

Personal PROBABILITY:
MOST OF LIFE'S EVENTS ARE *NOT REPEATABLE*. PERSONAL PROBABILITY IS AN *INDIVIDUAL'S PERSONAL ASSESSMENT* OF AN OUTCOME'S LIKELIHOOD. IF A GAMBLER BELIEVES THAT A HORSE HAS MORE THAN A 50% CHANCE OF WINNING, HE'LL TAKE AN EVEN BET ON THAT HORSE.

HOW DO YOU KNOW?

DA WISDOM OF DA TRACK...

AN *OBJECTIVIST* USES EITHER THE CLASSICAL OR FREQUENCY DEFINITION OF PROBABILITY. A *SUBJECTIVIST* OR *BAYESIAN* APPLIES FORMAL LAWS OF CHANCE TO HIS OWN, OR YOUR, PERSONAL PROBABILITIES.

HOW DO YOU KNOW THE ELEMENTARY OUTCOMES ARE EQUALLY LIKELY WITHOUT ROLLING THE DICE A BILLION TIMES?

WANNA BET?

OBJECTIVIST   BAYESIAN

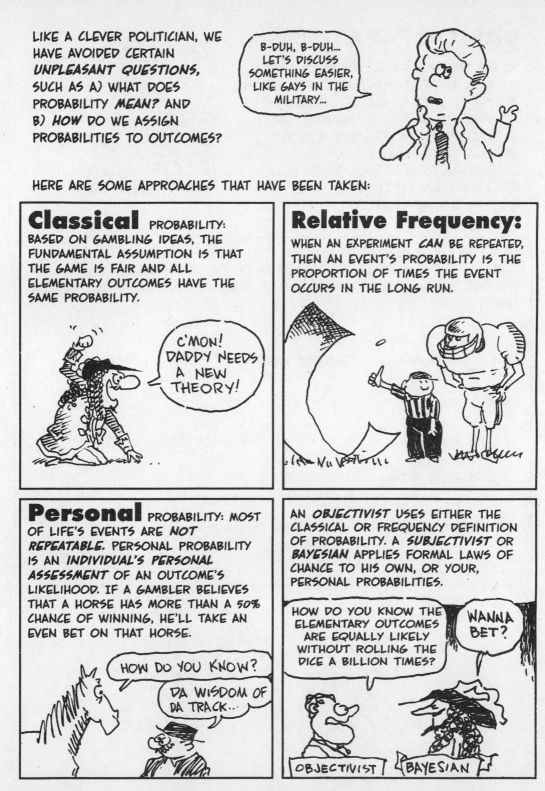

BASIC OPERATIONS

SO FAR, WE HAVE DISCUSSED ONLY THE PROBABILITY OF ELEMENTARY OUTCOMES. IN THEORY, THAT WOULD BE ENOUGH TO DESCRIBE ANY RANDOM EXPERIMENT, BUT IN PRACTICE IT'S PRETTY UNWIELDY. FOR EXAMPLE, EVEN SUCH AN ORDINARY OCCURRENCE AS ROLLING A SEVEN IS NOT AN ELEMENTARY OUTCOME... SO WE INTRODUCE A *NEW IDEA:*

AN *EVENT* IS A SET OF ELEMENTARY OUTCOMES. THE PROBABILITY OF AN EVENT IS THE SUM OF THE PROBABILITIES OF THE ELEMENTARY OUTCOMES IN THE SET. FOR INSTANCE, SOME EVENTS IN THE LIFE OF A TWO-DICED ROLLER ARE:

EVENT DESCRIPTION	EVENT'S ELEMENTARY OUTCOMES	PROBABILITY
A: DICE ADD TO 3	$\{(1,2), (2,1)\}$	$P(A) = \frac{2}{36}$
B: DICE ADD TO 6	$\{(1,5), (2,4), (3,3), (4,2), (5,1)\}$	$P(B) = \frac{5}{36}$
C: WHITE DIE SHOWS 1	$\{(1,1), (1,2), (1,3), (1,4), (1,5), (1,6)\}$	$P(C) = \frac{6}{36}$
D: BLACK DIE SHOWS 1	$\{(1,1), (2,1), (3,1), (4,1), (5,1), (6,1)\}$	$P(D) = \frac{6}{36}$

AND WHEN DO I GET MY SHIRT BACK?

THE BEAUTY OF USING EVENTS, RATHER THAN ELEMENTARY OUTCOMES, IS THAT WE CAN *COMBINE* EVENTS TO MAKE OTHER EVENTS, USING *LOGICAL OPERATIONS.* THE RELEVANT WORDS ARE *AND, OR,* AND *NOT.*

THAT IS, GIVEN EVENTS E AND F, WE CAN MAKE NEW EVENTS:

E **and** F :   THE EVENT E AND THE EVENT F BOTH OCCUR.

E **or** F :   THE EVENT E OR THE EVENT F OCCURS (OR BOTH DO).

not E :   THE EVENT E DOES NOT OCCUR.

COMBINING OUR PRIMITIVE DEFINITIONS OF PROBABILITY WITH THESE LOGICAL OPERATIONS WILL GIVE US SOME POWERFUL FORMULAS FOR MANIPULATING PROBABILITIES.

LET'S RETURN TO THE DICE-THROWING EXAMPLE. IF C IS THE EVENT, WHITE DIE = 1, AND D IS THE EVENT, BLACK DIE = 1, THEN:

C OR D IS THE ENTIRE SHADED AREA (WHERE ONE DIE OR THE OTHER IS 1).

C AND D IS WHERE THE SHADED AREAS OVERLAP (BOTH DICE ARE 1).

THIS ILLUSTRATES THE **ADDITION RULE:** FOR ANY EVENTS E, F,

$$P(E \text{ OR } F) = P(E) + P(F) - P(E \text{ AND } F)$$

ADDING P(E) + P(F) **DOUBLE COUNTS** THE ELEMENTARY OUTCOMES SHARED BY E AND F, SO WE HAVE TO SUBTRACT THE EXTRA AMOUNT, WHICH IS P(E AND F).

IN THE ABOVE EXAMPLE,

$$P(C \text{ OR } D) = \frac{11}{36}$$

AS YOU CAN SEE BY COUNTING ELEMENTARY OUTCOMES. LIKEWISE,

$$P(C \text{ AND } D) = \frac{1}{36}$$

AND WE CONFIRM THE FORMULA:

$$P(C) + P(D) - P(C \text{ AND } D)$$
$$= \frac{6}{36} + \frac{6}{36} - \frac{1}{36} = \frac{11}{36}$$
$$= P(C \text{ OR } D)$$

I AM HAVING A GLIMMER OF HOPE!

38

SOMETIMES, THE OVERLAP **E AND F** IS EMPTY, AND THE TWO EVENTS HAVE NO ELEMENTARY OUTCOMES IN COMMON. IN THAT CASE, WE SAY E AND F ARE *MUTUALLY EXCLUSIVE*, MAKING P(E AND F) = 0. HERE WE SEE THE MUTUALLY EXCLUSIVE EVENTS **A**, THE DICE ADD TO 3, AND **B**, THE DICE ADD TO 6.

FOR MUTUALLY EXCLUSIVE EVENTS, WE GET A *SPECIAL ADDITION RULE:* IF E AND F ARE MUTUALLY EXCLUSIVE, THEN

P(E OR F) = P(E) + P(F)

AND WE CHECK THAT $P(A \text{ OR } B) = \frac{7}{36} = \frac{2}{36} + \frac{5}{36} = P(A) + P(B)$

AND FINALLY, A *SUBTRACTION RULE:* FOR ANY EVENT E,

P(E) = 1 – P(NOT E)

THIS IS USEFUL WHEN P(NOT E) IS EASIER TO COMPUTE THAN P(E). FOR INSTANCE, LET **E** BE THE EVENT, A DOUBLE-1 IS *NOT THROWN.* THE EVENT NOT-E, A DOUBLE-1 *IS* THROWN, HAS PROBABILITY $P(\text{NOT } E) = \frac{1}{36}$.

SO

P(E) = 1–P(NOT E)

$= 1 - \frac{1}{36}$

$= \frac{35}{36}$

CAN WE SOLVE MY PROBLEM NOW? IT'S COLD...

THE FORMULAS WE JUST DERIVED ARE, IN FACT, ADEQUATE FOR ANSWERING DE MERE'S QUESTION— BUT NOT EASILY! (YOU MIGHT TRY USING THEM ON A SIMPLER QUESTION: WHAT'S THE PROBABILITY OF ROLLING AT LEAST ONE SIX IN TWO ROLLS OF A SINGLE DIE?) WE NEED *MORE MACHINERY!*

SO WE INTRODUCE

conditional probability

(AN ESSENTIAL CONCEPT IN STATISTICS!)

WHOA! LOOKS HEAVY!

SUPPOSE WE ALTER OUR EXPERIMENT SLIGHTLY, AND THROW THE WHITE DIE *BEFORE* THE BLACK DIE. WHAT'S THE PROBABILITY THAT THE FACES SUM TO 3?

BEFORE THE DICE ARE THROWN, THE PROBABILITY IS $P(A) = \frac{2}{36}$

NOW SUPPOSE THE WHITE DIE COMES UP 1 (EVENT C). WHAT'S THE PROBABILITY OF **A** NOW?

WE CALL IT THE **CONDITIONAL PROBABILITY** THAT EVENT A WILL OCCUR, GIVEN THE **CONDITION** THAT EVENT C HAS ALREADY OCCURRED. WE WRITE

$$P(A \mid C)$$

AND SAY "THE PROBABILITY OF A, **GIVEN C.**"

C

BEFORE ANY DICE WERE THROWN, THE SAMPLE SPACE HAD 36 OUTCOMES, BUT NOW THAT THE EVENT C HAS OCCURRED, THE OUTCOME MUST BELONG TO THE **REDUCED SAMPLE SPACE C.**

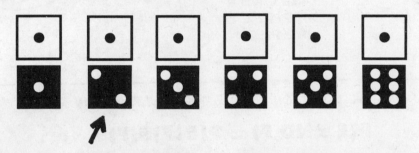

IN THE REDUCED SAMPLE SPACE OF SIX ELEMENTARY OUTCOMES, ONLY ONE OUTCOME (1,2) SUMS TO 3. SO THE CONDITIONAL PROBABILITY IS 1/6.

SEE HOW PROBABILITIES CHANGE AS THE WORLD EVOLVES?

MY SHIRT.

IN GENERAL, TO FIND THE CONDITIONAL PROBABILITY P(E|F), WE LOOK AT THE EVENT E AND F AS PART OF THE REDUCED SAMPLE SPACE F.

WE TRANSLATE THIS INTO A FORMAL DEFINITION: THE *CONDITIONAL PROBABILITY OF E, GIVEN F,* IS

$$P(E|F) = \frac{P(E \text{ and } F)}{P(F)}$$

WITH THE DICE, IT'S

$$\frac{P(A \text{ AND } C)}{P(C)} = \frac{\frac{1}{36}}{\frac{1}{6}}$$

$$= \frac{1}{6}$$

FROM WHICH YOU CAN DIRECTLY VERIFY SOME INTUITIVE FACTS:

$$P(E|E) = 1$$ (ONCE E OCCURS, IT'S CERTAIN.)

WHEN E AND F ARE MUTUALLY EXCLUSIVE,

$$P(E|F) = 0$$ (ONCE F HAS OCCURRED, E IS IMPOSSIBLE.)

REARRANGING THE DEFINITION GIVES US THE *MULTIPLICATION RULE:*

$$P(E \text{ AND } F) = P(E|F)P(F)$$

WHICH WE WOULD LIKE TO REDUCE TO A "SPECIAL" MULTIPLICATION RULE, UNDER THE FAVORABLE CIRCUMSTANCES THAT $P(E|F) = P(E)$. THAT WOULD BE EXCELLENT!

AND WHILE YOU'RE WAITING FOR THE NEXT PAGE, NOTE THAT SWAPPING E AND F PROVES THAT $P(F)P(E|F) = P(E)P(F|E)$.

INDEPENDENCE and the special multiplication rule.

TWO EVENTS E AND F ARE *INDEPENDENT* OF EACH OTHER IF THE OCCURRENCE OF ONE HAS *NO INFLUENCE* ON THE PROBABILITY OF THE OTHER. FOR INSTANCE, THE ROLL OF ONE DIE HAS NO EFFECT ON THE ROLL OF ANOTHER (UNLESS THEY'RE GLUED TOGETHER, MAGNETIC, ETC.!).

OOP!

IN TERMS OF CONDITIONAL PROBABILITY, THIS AMOUNTS TO SAYING $P(E) = P(E|F)$ OR, EQUIVALENTLY, $P(F) = P(F|E)$. WHEN E AND F ARE INDEPENDENT, WE GET A *SPECIAL MULTIPLICATION RULE:*

$$P(E \text{ AND } F) = P(E)P(F)$$

LET'S VERIFY THE INDEPENDENCE OF DICE, USING THE FORMULAS. C IS THE EVENT *WHITE DIE COMES UP 1;* D IS THE EVENT *BLACK DIE COMES UP 1,* AND WE HAVE:

$$P(C|D) = \frac{P(C \text{ AND } D)}{P(D)} = \frac{\frac{1}{36}}{\frac{1}{6}} = \frac{1}{6} = P(C)$$

BUT THE WHITE DIE SHOWING 1 OBVIOUSLY DOES AFFECT THE CHANCES THAT THE SUM OF THE TWO DICE IS 3!

$$P(A|C) = \frac{P(A \text{ AND } C)}{P(C)} = \frac{P(1,2)}{P(C)} = \frac{\frac{1}{36}}{\frac{1}{6}} = \frac{1}{6} \neq P(A) = \frac{1}{18}$$

SO THESE TWO EVENTS ARE *NOT INDEPENDENT.*

BEFORE GOING ON, LET'S SUMMARIZE ALL THE RULES WE'VE ACCUMULATED:

ADDITION RULE:

$$P(E \text{ OR } F) = P(E) + P(F) - P(E \text{ AND } F)$$

SPECIAL ADDITION RULE: WHEN E AND F ARE MUTUALLY EXCLUSIVE,

$$P(E \text{ OR } F) = P(E) + P(F)$$

SUBTRACTION RULE:

$$P(E) = 1 - P(\text{NOT } E)$$

MULTIPLICATION RULE:

$$P(E \text{ AND } F) = P(E | F)P(F)$$

SPECIAL MULTIPLICATION RULE: WHEN E AND F ARE INDEPENDENT,

$$P(E \text{ AND } F) = P(E)P(F)$$

AH, RULES! TO SAVE US FROM WASTEFUL THINKING!

AND NOW, DE MERE'S PROBLEM AT LAST... LET E BE THE EVENT OF GETTING AT LEAST ONE SIX IN FOUR ROLLS OF A SINGLE DIE. WHAT'S P(E)? THIS IS ONE OF THOSE EVENTS WHOSE NEGATIVE IS EASIER TO DESCRIBE: *NOT E* IS THE EVENT OF *GETTING NO SIXES IN FOUR THROWS.*

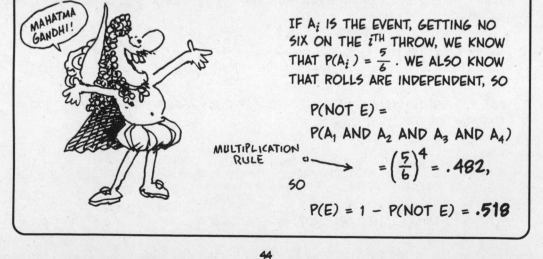

MAHATMA GANDHI!

IF A_i IS THE EVENT, GETTING NO SIX ON THE i^{TH} THROW, WE KNOW THAT $P(A_i) = \frac{5}{6}$. WE ALSO KNOW THAT ROLLS ARE INDEPENDENT, SO

$$P(\text{NOT } E) =$$
$$P(A_1 \text{ AND } A_2 \text{ AND } A_3 \text{ AND } A_4)$$

MULTIPLICATION RULE →

$$= \left(\frac{5}{6}\right)^4 = .482,$$

SO

$$P(E) = 1 - P(\text{NOT } E) = .518$$

44

NOW THE SECOND HALF: LET **F** BE THE EVENT, GETTING AT LEAST ONE DOUBLE SIX IN 24 THROWS. AGAIN, **NOT F** IS EASIER TO DESCRIBE. IT'S THE EVENT OF GETTING NO DOUBLE SIXES.

BRAVO! I CAN DIE HAPPY!!

DE MERE

IF B_i IS THE EVENT, NO DOUBLE SIX IS THROWN ON THE i^{TH} ROLL, THEN NOT F = B_1 AND B_2 AND... B_{24}. THE PROBABILITY OF EACH B IS

$$P(B_i) = \frac{35}{36} \text{ , SO}$$

$$P(\text{NOT F}) = \left(\frac{35}{36}\right)^{24} = .509$$

(BY THE MULTIPLICATION RULE) AND WE CONCLUDE THAT

$$P(F) = 1 - P(\text{NOT F}) = 1 - .509$$

$$= .491$$

DE MERE TOLD PASCAL HE HAD ACTUALLY OBSERVED THAT EVENT F OCCURRED LESS OFTEN THAN EVENT E, BUT HE WAS AT A LOSS TO EXPLAIN WHY... FROM WHICH WE CONCLUDE THAT DE MERE GAMBLED OFTEN AND *KEPT CAREFUL RECORDS!!*

WHAT ARE MY ODDS OF GETTING IN?

NOW LET'S LEAVE THE CASINO AND REJOIN THE REAL WORLD...

BAYES THEOREM and the
case of the false positives...

FOR A MORE SERIOUS APPLICATION OF
CONDITIONAL PROBABILITY, LET'S ENTER
AN ARENA OF LIFE AND DEATH...

SUPPOSE A RARE DISEASE INFECTS ONE OUT OF EVERY 1000 PEOPLE IN A
POPULATION...

AND SUPPOSE THAT THERE IS A GOOD, BUT NOT PERFECT, TEST FOR THIS
DISEASE: IF A PERSON HAS THE DISEASE, THE TEST COMES BACK POSITIVE **99%**
OF THE TIME. ON THE OTHER HAND, THE TEST ALSO PRODUCES SOME *FALSE
POSITIVES*. ABOUT **2%** OF UNINFECTED PATIENTS ALSO TEST POSITIVE. AND YOU
JUST TESTED POSITIVE. WHAT ARE YOUR CHANCES OF HAVING THE DISEASE?

LET'S PUT
IT THIS WAY:
SHOULD I
PAY IN ADVANCE?

WE HAVE TWO EVENTS TO WORK WITH:

A : PATIENT HAS THE DISEASE
B : PATIENT TESTS POSITIVE.

THE INFORMATION ABOUT THE TEST'S
EFFECTIVENESS CAN BE WRITTEN

$P(A) = .001$ (ONE PATIENT IN 1000 HAS THE DISEASE)

$P(B|A) = .99$ (PROBABILITY OF A POSITIVE TEST, GIVEN INFECTION, IS .99)

$P(B|NOT\ A) = .02$ (PROBABILITY OF A FALSE POSITIVE, GIVEN NO INFECTION, IS .02)

AND WE ASK

$P(A|B) = $ WHAT?   (PROBABILITY OF HAVING THE DISEASE, GIVEN A POSITIVE TEST)

SINCE THE TREATMENT FOR THIS DISEASE HAS SERIOUS SIDE EFFECTS, THE DOCTOR, HER LAWYER, AND HER LAWYER'S LAWYER CALL ON JOE BAYES, CP (CONSULTING PROBABILIST), FOR AN ANSWER. JOE DERIVES A THEOREM FIRST PROVED BY HIS ANCESTOR, THE REV. THOMAS *BAYES* (1744-1809).

JOE BEGINS WITH A 2X2 TABLE, WHICH DIVIDES THE SAMPLE SPACE INTO FOUR MUTUALLY EXCLUSIVE EVENTS. IT DISPLAYS EVERY POSSIBLE COMBINATION OF DISEASE STATE AND TEST RESULT.

	A	NOT A
B	A AND B	NOT A AND B
NOT B	A AND NOT B	NOT A AND NOT B

LET'S FIND THE PROBABILITIES OF EACH EVENT IN THE TABLE:

	A	NOT A	SUM
B	P(A AND B)	P(NOT A AND B)	P(B)
NOT B	P(A AND NOT B)	P(NOT A AND NOT B)	P(NOT B)
	P(A)	P(NOT A)	1

THE PROBABILITIES IN THE MARGINS ARE FOUND BY SUMMING ACROSS ROWS AND DOWN COLUMNS.

BY DEFINITION!

NOW COMPUTE:

$$P(A \text{ AND } B) = P(B|A)P(A) = (.99)(.001) = .00099$$

$$P(\text{NOT } A \text{ AND } B) = P(B|\text{NOT } A)P(\text{NOT } A) = (.02)(.999) = .01998$$

ALLOWING US TO FILL IN SOME ENTRIES:

	A	NOT A	SUM
B	.00099	.01998	.02097
NOT B	P(A AND NOT B)	P(NOT A AND NOT B)	P(NOT B)
	.001	.999	1

WE FIND THE REMAINING PROBABILITIES BY SUBTRACTING IN THE COLUMNS, THEN ADDING ACROSS THE ROWS.

THE FINAL TABLE IS:

	A	NOT A		
B	.00099	.01998	.02097	P(B)
NOT B	.00001	.97902	.97903	P(NOT B)
	.001	.999	1	
	P(A)	P(NOT A)		

FROM WHICH WE DIRECTLY DERIVE

$$P(A|B) = \frac{P(A \text{ AND } B)}{P(B)} = \frac{.00099}{.02097} = .0472$$

DESPITE THE HIGH ACCURACY OF THE TEST, *LESS THAN 5%* OF THOSE WHO TEST POSITIVE ACTUALLY HAVE THE DISEASE! THIS IS CALLED THE *FALSE POSITIVE PARADOX.*

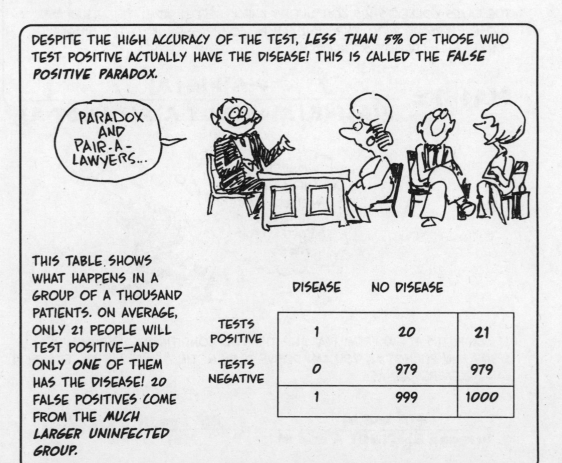

PARADOX AND PAIR-A-LAWYERS...

THIS TABLE SHOWS WHAT HAPPENS IN A GROUP OF A THOUSAND PATIENTS. ON AVERAGE, ONLY 21 PEOPLE WILL TEST POSITIVE—AND ONLY *ONE* OF THEM HAS THE DISEASE! 20 FALSE POSITIVES COME FROM THE *MUCH LARGER UNINFECTED GROUP.*

	DISEASE	NO DISEASE	
TESTS POSITIVE	1	20	21
TESTS NEGATIVE	0	979	979
	1	999	1000

WHAT'S THE PHYSICIAN TO DO? JOE BAYES ADVISES HER NOT TO START
TREATMENT ON THE BASIS OF THIS TEST ALONE. THE TEST DOES PROVIDE
INFORMATION, HOWEVER: WITH A POSITIVE TEST THE PATIENT'S CHANCE OF
HAVING THE DISEASE INCREASED FROM 1 IN 1000 TO 1 IN 23. THE DOCTOR
FOLLOWS UP WITH MORE TESTS.

JOE BAYES COLLECTS HIS CONSULTING CHECK BEFORE ADMITTING THAT ALL
THOSE STEPS HE WENT THROUGH CAN BE COMPRESSED INTO THE SINGLE
FORMULA CALLED BAYES THEOREM:

$$P(A|B) = \frac{P(A)P(B|A)}{P(A)P(B|A) + P(NOT\ A)P(B|NOT\ A)}$$

IT COMPUTES $P(A|B)$ FROM $P(A)$ AND THE TWO CONDITIONAL PROBABILITIES
$P(B|A)$ AND $P(B|NOT\ A)$. YOU CAN DERIVE IT BY NOTING THAT THE BIG FRACTION
CAN BE EXPRESSED AS

$$\frac{P(A\ and\ B)}{P(A\ and\ B) + P(NOT\ A\ and\ B)} = \frac{P(A\ and\ B)}{P(B)} = P(A|B)$$

IN THIS CHAPTER, WE COVERED THE BASICS OF PROBABILITY: ITS DEFINITION, SAMPLE SPACES AND ELEMENTARY OUTCOMES, CONDITIONAL PROBABILITY, AND SOME BASIC FORMULAS FOR COMPUTING PROBABILITIES. WE ILLUSTRATED THESE IDEAS USING A 2-DICE SAMPLE SPACE. FOR THE MODERN GAMBLER, PROBABILITY IS THE POWER TOOL OF CHOICE.

AND FINALLY, IN THE MEDICAL EXAMPLE, WE SHOWED HOW THESE ABSTRACT IDEAS COULD HELP TO MAKE GOOD DECISIONS IN THE FACE OF *IMPERFECT INFORMATION* AND *REAL RISKS*—THE *ULTIMATE GOAL* OF STATISTICS.

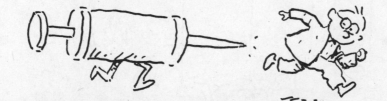

BUT THIS IS JUST THE BEGINNING. FOR US, PROBABILITY IS ONLY A *TOOL*—AN ESSENTIAL TOOL, TO BE SURE—IN THE STUDY OF STATISTICS. IN THE CHAPTERS THAT FOLLOW, WE'LL EXPLORE THE SUBTLE RELATIONSHIP BETWEEN PROBABILITY, VARIATIONS IN STATISTICAL DATA, AND OUR CONFIDENCE IN INTERPRETING THE MEANING OF OUR OBSERVATIONS.

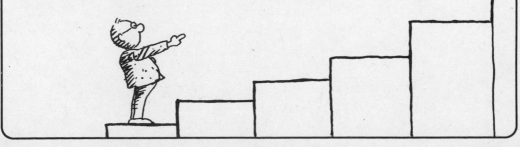

◆ Chapter 4 ◆
RANDOM VARIABLES

IN CHAPTER 2, WE SAW THAT OBSERVATIONS OF NUMERICAL
DATA, LIKE STUDENTS' WEIGHTS, CAN BE GRAPHED AND
SUMMARIZED IN TERMS OF MIDPOINTS, SPREADS, OUTLIERS, ETC.
IN CHAPTER 3, WE SAW HOW PROBABILITIES CAN BE ASSIGNED
TO THE OUTCOMES OF A RANDOM EXPERIMENT.

IF WE IMAGINE A RANDOM EXPERIMENT REPEATED MANY TIMES,
WE EXPECT THAT THE ACTUAL OUTCOMES OVER TIME WILL BE
GOVERNED BY THEIR PROBABILITIES. THE PROBABILITIES FORM A
MODEL FOR REAL-LIFE EXPERIMENTS... SO WHY NOT DO FOR THE
MODEL WHAT WE'VE ALREADY DONE FOR THE DATA IT DESCRIBES?

THE KEY IDEA IS THE *RANDOM VARIABLE*, WHICH WE WRITE AS A LARGE

X

A RANDOM VARIABLE IS DEFINED AS THE *NUMERICAL OUTCOME OF A RANDOM EXPERIMENT.*

FOR EXAMPLE, IMAGINE DRAWING ONE STUDENT AT RANDOM FROM THE STUDENT BODY. THAT'S THE RANDOM EXPERIMENT. THE STUDENT'S *HEIGHT, WEIGHT, FAMILY INCOME, S.A.T. SCORE,* AND *GRADE POINT AVERAGE* ARE ALL NUMERICAL *VARIABLES* DESCRIBING PROPERTIES OF THE RANDOMLY SELECTED STUDENT. THEY'RE ALL *RANDOM VARIABLES.*

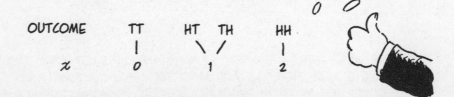

THE ADMINISTRATION'S JOB IS TO TURN STUDENTS INTO STATISTICS!

ANOTHER EXAMPLE: TOSS TWO COINS (THE RANDOM EXPERIMENT) AND RECORD THE *NUMBER* OF HEADS: 0, 1, OR 2.

OUTCOME	TT	HT TH	HH
x	0	1	2

NOTE THE NOTATION! THE VARIABLE IS WRITTEN WITH A CAPITAL X. THE LOWERCASE x REPRESENTS A SINGLE VALUE OF X, FOR EXAMPLE $x = 2$, IF HEADS COMES UP TWICE.

ANOTHER EXAMPLE IS BASED ON THE FAMILIAR TOSS OF TWO DICE. LET Y REPRESENT THE SUM OF THE DOTS ON THE TWO DICE. FOR THIS RANDOM VARIABLE, Y CAN BE ANY NUMBER BETWEEN 2 AND 12.

$$Y = 7$$

NOW WE WANT TO LOOK AT THE *PROBABILITIES* OF THE OUTCOMES. FOR THE PROBABILITY THAT THE RANDOM VARIABLE X HAS THE VALUE x, WE WRITE $Pr(X = x)$, OR JUST $p(x)$. FOR THE COIN-FLIPPING RANDOM VARIABLE X, WE CAN MAKE THE TABLE:

x	0	1	2
$Pr(X=x)$	$\frac{1}{4}$	$\frac{1}{2}$	$\frac{1}{4}$

THIS TABLE IS CALLED THE *PROBABILITY DISTRIBUTION* OF THE RANDOM VARIABLE X.

FOR THE RANDOM VARIABLE Y (THE SUM OF TWO DICE), THE PROBABILITY DISTRIBUTION LOOKS LIKE THIS:

y	2	3	4	5	6	7	8	9	10	11	12
$Pr(Y=y)$	$\frac{1}{36}$	$\frac{2}{36}$	$\frac{3}{36}$	$\frac{4}{36}$	$\frac{5}{36}$	$\frac{6}{36}$	$\frac{5}{36}$	$\frac{4}{36}$	$\frac{3}{36}$	$\frac{2}{36}$	$\frac{1}{36}$

YUP! THAT'S WHY I GAVE UP DICIN'!

NOW LET'S DRAW GRAPHS, OR **HISTOGRAMS**, SHOWING THESE
PROBABILITY DISTRIBUTIONS. FOR EACH VALUE OF X, WE DRAW A BAR
EQUAL IN HEIGHT TO $p(x)$.

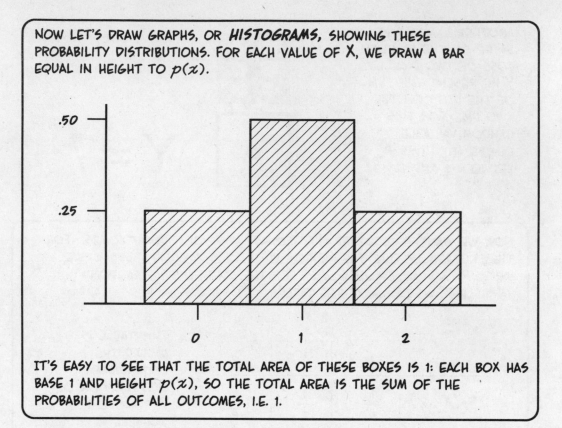

IT'S EASY TO SEE THAT THE TOTAL AREA OF THESE BOXES IS 1: EACH BOX HAS
BASE 1 AND HEIGHT $p(x)$, SO THE TOTAL AREA IS THE SUM OF THE
PROBABILITIES OF ALL OUTCOMES, I.E. 1.

HERE'S THE PROBABILITY HISTOGRAM OF THE RANDOM VARIABLE Y, SHOWING
THE PROBABILITY DISTRIBUTION OF THE SUM OF TWO DICE:

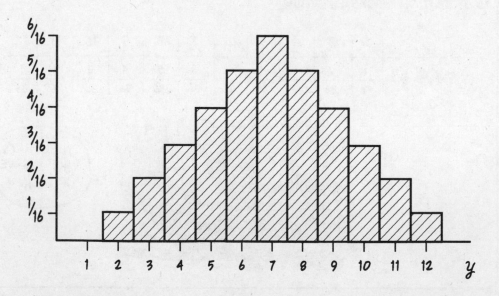

WHY DO WE CALL THESE GRAPHS HISTOGRAMS? YOU'LL RECALL THAT IN CHAPTER 2, A HISTOGRAM WAS A GRAPH THAT DISPLAYED HOW MANY DATA POINTS LAY IN EACH OF A SERIES OF INTERVALS:

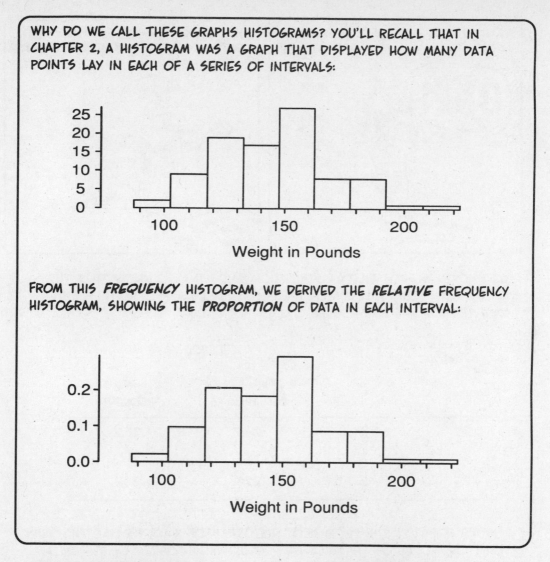

FROM THIS *FREQUENCY* HISTOGRAM, WE DERIVED THE *RELATIVE* FREQUENCY HISTOGRAM, SHOWING THE *PROPORTION* OF DATA IN EACH INTERVAL:

BUT YOU'LL RECALL THAT, BY ONE DEFINITION, PROBABILITY IS THE *RELATIVE FREQUENCY OF AN EVENT "IN THE LONG RUN."* IF WE REPEAT THE RANDOM EXPERIMENT MANY TIMES, THE *RELATIVE FREQUENCY* HISTOGRAM OF THE OUTCOMES SHOULD COME TO LOOK VERY MUCH LIKE THE RANDOM VARIABLE'S *PROBABILITY* HISTOGRAM!

WE KNOW X'S PROBABILITY DISTRIBUTION, AND WE ALSO KNOW THAT THE ACTUAL COIN FLIPS WILL MATCH THE PROBABILITIES APPROXIMATELY. AFTER 1000 TOSSES, THE MAD TOSSER TALLIES HER DATA:

PROBABILITY MODEL		OBSERVED DATA	
$p(x)$	x	n_x = NUMBER OF OCCURRENCES	$\frac{n_x}{n}$ = RELATIVE FREQUENCY
.25	0	260	.260
.5	1	517	.517
.25	2	223	.223

AND WE SEE THAT THE PROBABILITY HISTOGRAM OF X LOOKS LIKE THE "PURE FORM" OR MODEL OF THE RELATIVE FREQUENCY HISTOGRAM OF THE DATA.

58

TO EXTEND THE ANALOGY BETWEEN RELATIVE FREQUENCY AND DATA, WE SHOULD NOW BE WILLING TO TALK ABOUT THE MEAN AND VARIANCE (OR STANDARD DEVIATION) OF A PROBABILITY DISTRIBUTION...

LOVE THOSE ABSTRAC-TIONS!

AND JUST TO REMIND OURSELVES THAT WE'RE IN THE REALM OF THE ABSTRACT, WE BREAK OUT SOME *GREEK LETTERS*...

MEAN AND VARIANCE OF RANDOM VARIABLES

WE USE SPECIAL TERMINOLOGY AND SYMBOLS TO DISTINGUISH BETWEEN THE PROPERTIES OF DATA SETS AND PROBABILITY DISTRIBUTIONS:

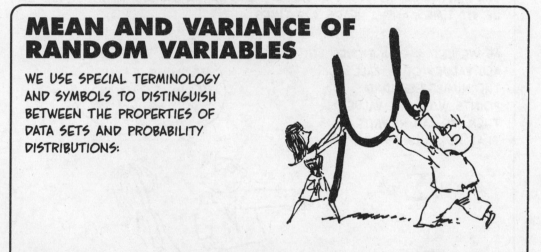

PROPERTIES OF DATA ARE CALLED *SAMPLE* PROPERTIES, WHILE PROPERTIES OF THE PROBABILITY DISTRIBUTION ARE CALLED *MODEL* OR *POPULATION* PROPERTIES. WE USE THE GREEK LETTER μ (MU) FOR THE POPULATION MEAN, AND σ (LOWERCASE SIGMA) FOR THE POPULATION STANDARD DEVIATION. (FOR DATA, WE USE THE ROMAN SYMBOLS $\bar{x}$ AND s.)

BECAUSE ROMANS WERE SHORT ON THEORY AND LONG ON CEMENT, AND STUFF LIKE THAT...

THE SAMPLE MEAN WAS DEFINED
BY THE EQUATION

$$\bar{x} = \frac{1}{n} \sum_{i=1}^{n} x_i$$

NOW SOME OF THESE DATA POINTS x_i MAY WELL HAVE EQUAL VALUES. THINK
OF THE MAD COIN TOSSER: THE ONLY AVAILABLE VALUES WERE 0, 1, AND 2, AND
SHE MADE 1000 TOSSES. THE VALUE 0 WAS TAKEN ON 260 TIMES, 1 HEAD CAME
UP 517 TIMES, AND 2 HEADS, 223 TIMES.

AS WE LET x RANGE OVER
ALL VALUES OF X, CALL n_x
THE NUMBER OF DATA
POINTS WITH THE VALUE x.
THEN WE CAN REWRITE
THAT FORMULA AS

$$\bar{x} = \frac{1}{n} \sum_{all\ x} n_x x$$

OR

$$\bar{x} = \sum_{all\ x} x \frac{n_x}{n}$$

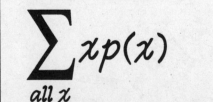

AH! BUT NOW $\frac{n_x}{n}$ IS THE RELATIVE FREQUENCY... THE "APPROXIMATE
PROBABILITY..." THE NUMBER THAT APPROACHES $p(x)$...SO, BY ANALOGY, WE
FORM THE EXPRESSION

$$\sum_{all\ x} x p(x)$$

AND DEFINE THAT AS THE
**MEAN OF THE PROBABILITY
DISTRIBUTION.**

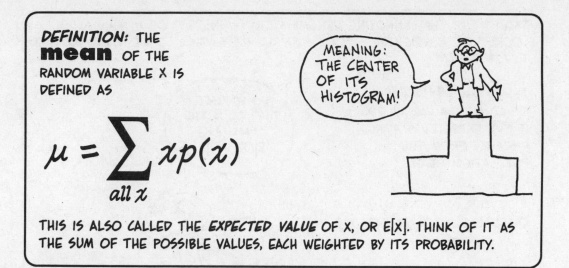

DEFINITION: THE **mean** OF THE RANDOM VARIABLE X IS DEFINED AS

$$\mu = \sum_{all\ x} x p(x)$$

MEANING: THE CENTER OF ITS HISTOGRAM!

THIS IS ALSO CALLED THE *EXPECTED VALUE* OF X, OR E[X]. THINK OF IT AS THE SUM OF THE POSSIBLE VALUES, EACH WEIGHTED BY ITS PROBABILITY.

THE MAD COIN TOSSER'S EXPERIMENT ALLOWS US TO COMPARE HER SAMPLE MEAN $\bar{x}$ WITH OUR MODEL MEAN μ:

	SAMPLE				MODEL	
x	$\frac{n_x}{n}$	$x\frac{n_x}{n}$		x	$p(x)$	$x p(x)$
0	.26	0		0	.25	0
1	.517	.517		1	.5	.5
2	.223	.446		2	.25	.5
		.963 = $\bar{x}$				1 = μ

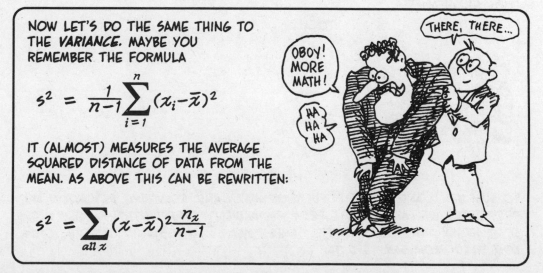

NOW LET'S DO THE SAME THING TO THE *VARIANCE*. MAYBE YOU REMEMBER THE FORMULA

$$s^2 = \frac{1}{n-1}\sum_{i=1}^{n}(x_i - \bar{x})^2$$

IT (ALMOST) MEASURES THE AVERAGE SQUARED DISTANCE OF DATA FROM THE MEAN. AS ABOVE THIS CAN BE REWRITTEN:

$$s^2 = \sum_{all\ x}(x - \bar{x})^2 \frac{n_x}{n-1}$$

OBOY! MORE MATH! HA HA HA

THERE, THERE...

EXCEPT FOR THAT ANNOYING DENOMINATOR n-1 INSTEAD OF n, THIS ALSO LOOKS LIKE A WEIGHTED SUM OF SQUARED DISTANCES... SO WE MAKE ANOTHER DEFINITION:

THE **variance**
OF A RANDOM VARIABLE X IS THE EXPECTED SQUARED DISTANCE FROM THE POPULATION MEAN:

DO YOU SEE THAT σ^2 IS THE SAME AS $E[(X-\mu)^2]$?

$$\sigma^2 = \sum_{all\ x} (x-\mu)^2 p(x)$$

THE **standard deviation** σ
IS THE SQUARE ROOT OF THE VARIANCE.

WE USE THE TABLE FROM THE LAST PAGE TO FIND THE VARIANCE OF THE TWO-COIN TOSS (FOR WHICH $\mu = 1$).

x	$p(x)$	$(x-\mu)^2 p(x)$
0	.25	$(0-1)^2 .25 = .25$
1	.5	$(1-1)^2 .50 = 0$
2	.25	$(2-1)^2 .25 = .25$
TOTAL		$.50 = \sigma^2$

TO SUM UP: μ AND σ, THE POPULATION *MEAN* AND *STANDARD DEVIATION*, ARE PROPERTIES WE CAN COMPUTE FROM *PROBABILITY DISTRIBUTIONS*. THEY ARE COMPLETELY ANALOGOUS TO THE SAMPLE MEAN $\bar{x}$ AND STANDARD DEVIATION s COMPUTED FROM SAMPLE DATA.

OUR EXAMPLES SO FAR HAVE BEEN *DISCRETE* RANDOM VARIABLES. THEIR OUTCOMES ARE A SET OF ISOLATED ("DISCRETE") VALUES, LIKE THOSE WE SAW IN CHAPTER 3, BUT THERE ARE ALSO

Continuous Random Variables

LET'S IMAGINE A RANDOM EXPERIMENT IN WHICH **ALL OUTCOMES HAVE PROBABILITY ZERO.** THAT'S RIGHT, $p(x) = 0$ FOR EVERY x.

A SIMPLE EXAMPLE IS A BALANCED, SPINNING POINTER. IT CAN STOP ANYWHERE IN THE CIRCLE. IF X REPRESENTS THE PROPORTION OF THE TOTAL CIRCUMFERENCE IT LANDS ON, THE RANDOM VARIABLE X CAN TAKE ON ANY VALUE BETWEEN 0 AND 1—AN *INFINITE* RANGE OF VALUES.

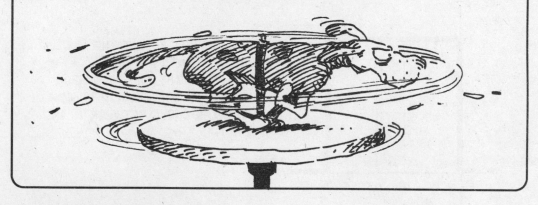

SOME PROBABILITIES ARE EASY TO FIND, LIKE THE PROBABILITY THAT X FALLS WITHIN A RANGE: FOR EXAMPLE, $Pr(.25 \leq X \leq .75) = .5$, BECAUSE IT'S HALF THE CIRCLE. BUT WHAT ABOUT $Pr(X = .5)$? SINCE X CAN TAKE ON AN INFINITE NUMBER OF VALUES, AND ALL OF THESE VALUES ARE EQUALLY LIKELY, THE PROBABILITY THAT X IS EXACTLY .5 (OR EXACTLY ANYTHING) IS PRECISELY 0.

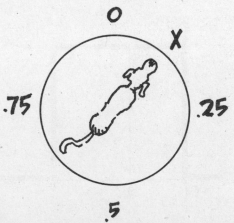

HOW CAN WE DRAW A PICTURE OF THIS?
BY ANALOGY WITH THE CASE OF
DISCRETE PROBABILITIES, WE TRY TO
SEE CONTINUOUS PROBABILITIES AS
AREAS UNDER SOMETHING. FOR THE
SPINNING POINTER, THE "SOMETHING"
LOOKS LIKE THIS:

$f(x) = 0$ WHEN $x < 0$

$f(x) = 1$ WHEN $0 \leq x \leq 1$

$f(x) = 0$ WHEN $x > 1$

THE PROBABILITY THAT THE
POINTER POINTS ANYWHERE
BETWEEN a AND b IS PRECISELY
THE AREA OF THE SHADED REGION
UNDER THE CURVE BETWEEN a AND
b (IN THIS CASE, $b-a$).

THE PROBABILITY OF AN EXACT
OUTCOME, HOWEVER, IS THE "AREA"
OVER A POINT, WHICH IS *ZERO.*
(AND NOTE THAT THE TOTAL AREA
UNDER THE CURVE IS EXACTLY 1.)

THE SAME PICTURE DESCRIBES THE *RANDOM NUMBER GENERATOR* FOUND ON MOST COMPUTERS AND SOME CALCULATORS. PRESS THE BUTTON; OUT POPS A NUMBER BETWEEN 0 AND 1; AND ALL THE NUMBERS ARE EQUALLY LIKELY, JUST AS WITH THE SPINNING POINTER.

HEY... THIS IS KIND OF FUN...

PUNCH PUNCH PUNCH

BUT SADLY, THEY AREN'T TRULY RANDOM. THEY'RE PRODUCED BY SOME ALGORITHM, SO, TO BE ACCURATE, WE CALL THEM *PSEUDO-RANDOM* NUMBERS.

THE CURVE $y = f(x)$ IN THIS EXAMPLE IS CALLED THE *PROBABILITY DENSITY* OF THE CONTINUOUS RANDOM VARIABLE X. EVERY CONTINUOUS RANDOM VARIABLE HAS ITS OWN DENSITY FUNCTION. THE PROBABILITY $Pr(a \leq X \leq b)$ IS THE AREA UNDER THE CURVE BETWEEN THE x-VALUES a AND b.

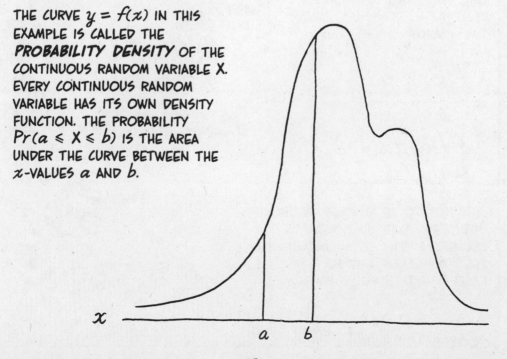

IN GENERAL, THE PROBABILITY
DENSITY WON'T BE SO SIMPLE,
AND COMPUTING THE AREAS CAN
BE FAR FROM TRIVIAL.

$$\int_a^b f(x)\,dx$$

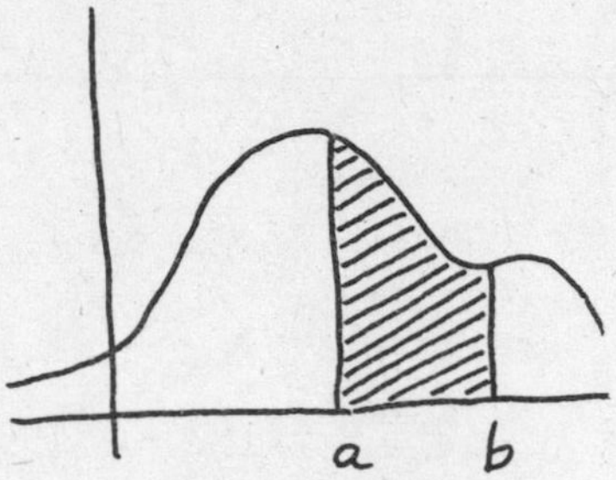

WE HAVE TO USE CALCULUS
NOTATION TO DESCRIBE THE
AREA UNDER THE CURVE $f(x)$.
THIS SYMBOL IS READ "THE
INTEGRAL OF f FROM a TO b."

GLEEP!

THERE, THERE...

LIKE DISCRETE PROBABILITIES,
CONTINUOUS DENSITIES HAVE
TWO FAMILIAR PROPERTIES:

GIBBER GIBBER

$$f(x) \geq 0$$

$$\int_{-\infty}^{\infty} f(x)\,dx = 1$$

(TRY NOT TO BE ALARMED BY THOSE
INFINITIES... THEY JUST MEAN WE'RE
LOOKING AT THE TOTAL AREA UNDER
THE CURVE FROM END TO END,
EXCEPT THAT THERE IS NO END!)

ALTHOUGH THE NOTATION MAY BE UNFAMILIAR, ALL IT MEANS IS AN *AREA*.. THE INTEGRAL SIGN ITSELF IS A STRETCHED "S," FOR SUM, WHICH THE INTEGRAL, IN SOME SENSE, IS.

AS A SUMLIKE SOMETHING, THE INTEGRAL SERVES TO DEFINE THE

MEAN AND VARIANCE of a continuous random variable.

$$\mu = \int_{-\infty}^{\infty} x f(x)\,dx$$

BY ANALOGY WITH THE DISCRETE FORMULAS:

$$\mu = \sum_{all\ x} x p(x)$$

$$\sigma^2 = \int_{-\infty}^{\infty} (x-\mu)^2 f(x)\,dx$$

$$\sigma^2 = \sum_{all\ x} (x-\mu)^2 p(x)$$

ALTHOUGH IT MAY NOT BE OBVIOUS FROM THE FORMULAS, THESE DEFINITIONS OF MEAN AND VARIANCE ARE ENTIRELY CONSISTENT WITH THEIR ROLE AS CENTER AND AVERAGE SPREAD OF THE PROBABILITIES GIVEN BY THE DENSITY $f(x)$. THE PICTURE TO KEEP IN MIND IS THIS:

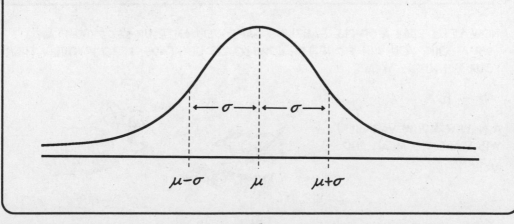

ADDING
random variables

ONCE YOU KNOW THE MEAN AND
VARIANCE OF A RANDOM VARIABLE,
WHAT CAN YOU DO WITH THEM?
WELL, FOR ONE THING, YOU CAN
FIND THE MEAN AND VARIANCE OF
SOME *OTHER* RANDOM VARIABLES...

OH...
THAT SOUNDS
USEFUL...

FOR EXAMPLE, LOOK AT A FAIR COIN TOSS. LET X = 1 IF THE COIN COMES UP
HEADS AND 0 IF IT COMES UP TAILS.

x	0	1
$p(x)$	.5	.5

NOTHING
NEW HERE..

BY NOW, YOU SHOULD BE ABLE
TO FIND THE MEAN

$$E[X] = 0 \cdot p(0) = 1 \cdot p(1)$$
$$= 0 + .5$$
$$= .5$$

AND THE VARIANCE

$$\sigma^2 = (0 - .5)^2 p(0) + (1 - .5)^2 p(1)$$
$$= .25$$

NOW LET'S PLAY A SIMPLE GAMBLING GAME: YOU ANTE UP $6.00 TO PLAY; I
FLIP A COIN; YOU WIN $10 IF THE COIN COMES UP HEADS, ZERO IF TAILS. THEN
YOUR WINNINGS **W** ARE

$$W = 10X - 6$$

A NEW RANDOM VARIABLE!
WHAT ARE ITS MEAN AND
VARIANCE?

A LITTLE THOUGHT SHOULD CONVINCE YOU THAT E[W] IS GIVEN BY

$$E[W] = E[10X - 6]$$
$$= 10E[X] - 6$$

WHICH WORKS OUT TO

$$10(0.5) - 6 = -1$$

YOU CAN CHECK IT USING THIS TABLE:

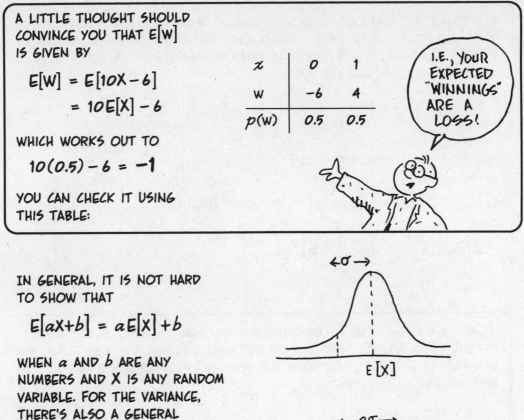

x	0	1
W	-6	4
$p(W)$	0.5	0.5

I.E., YOUR EXPECTED "WINNINGS" ARE A LOSS!

IN GENERAL, IT IS NOT HARD TO SHOW THAT

$$E[aX+b] = aE[X]+b$$

WHEN a AND b ARE ANY NUMBERS AND X IS ANY RANDOM VARIABLE. FOR THE VARIANCE, THERE'S ALSO A GENERAL RESULT:

$$\sigma^2(aX+b) = a^2\sigma^2(X)$$

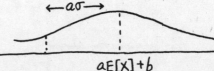

IN THE GAMBLING GAME ABOVE, THE POSSIBLE OUTCOMES ARE -6 AND 4, SO IT'S CLEAR THAT THE VARIANCE OF W MUST BE GREATER THAN THE VARIANCE OF X. IN FACT,

$$\sigma^2(W) = \sigma^2(10X+6)$$
$$= 100\,\sigma^2(X)$$
$$= 25$$

AND

$$\sigma(W) = 5$$

SOUNDS LIKE A SUCKER BET TO ME!

YOU CAN ALSO **ADD** TWO RANDOM VARIABLES TOGETHER. FOR INSTANCE, SUPPOSE WE TOSS A COIN **TWICE**. THE NUMBER OF HEADS ON BOTH TOSSES IS $X_1 + X_2$, WHERE X_1 AND X_2 ARE THE RANDOM VARIABLES GIVING THE RESULTS OF THE FIRST AND SECOND TOSSES.

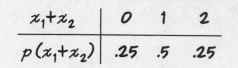

$x_1 + x_2$	0	1	2
$p(x_1 + x_2)$	.25	.5	.25

AGAIN, IT'S EASY TO SEE THAT

$$E[X_1 + X_2] = E[X_1] + E[X_2]$$

(DON'T ASK ABOUT THE PROBABILITY DISTRIBUTION OF $X_1 + X_2$, BECAUSE IT DEPENDS IN A COMPLICATED WAY ON THE TWO ORIGINAL DISTRIBUTIONS. FOR EXAMPLE, IF X_1 AND X_2 ARE BOTH THE SPINNING POINTER DISTRIBUTION, THE HISTOGRAMS ACT LIKE THIS:)

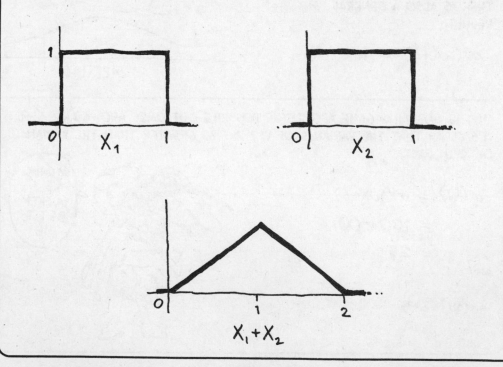

THE VARIANCE OF THE SUM OF RANDOM VARIABLES HAS A SIMPLE FORM IN THE SPECIAL CASE WHEN THE VARIABLES X AND Y ARE **INDEPENDENT.** THE TECHNICAL DEFINITION OF INDEPENDENCE IS BASED ON THE PROBABILITY PROPERTY P(A AND B) = P(A)P(B)... BUT FOR US, INDEPENDENCE JUST MEANS THAT X AND Y ARE GENERATED BY **INDEPENDENT MECHANISMS,** SUCH AS FLIPS OF A COIN, ROLLS OF A DIE, ETC.

OUTSIDE THE CASINO, IT'S HARD TO FIND COMPLETE INDEPENDENCE...

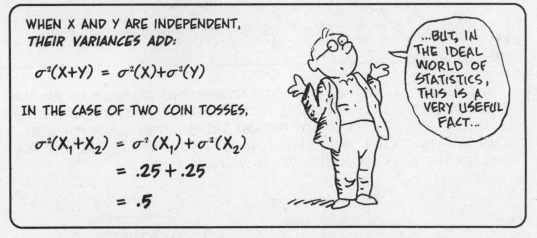

WHEN X AND Y ARE INDEPENDENT, **THEIR VARIANCES ADD:**

$$\sigma^2(X+Y) = \sigma^2(X) + \sigma^2(Y)$$

IN THE CASE OF TWO COIN TOSSES,

$$\sigma^2(X_1+X_2) = \sigma^2(X_1) + \sigma^2(X_2)$$
$$= .25 + .25$$
$$= .5$$

...BUT, IN THE IDEAL WORLD OF STATISTICS, THIS IS A VERY USEFUL FACT...

ALL OF THIS CAN BE GENERALIZED TO THE SUM OF MANY RANDOM VARIABLES:

$$E\left[\sum_{i=1}^{n} X_i\right] = \sum_{i=1}^{n} E[X_i]$$

AND, WHEN THE X_i ARE ALL INDEPENDENT,

$$\sigma^2\left(\sum_{i=1}^{n} X_i\right) = \sum_{i=1}^{n} \sigma^2(X_i)$$

THESE CALCULATIONS LIE AT THE HEART OF MOST SAMPLING THEORY AND STATISTICS. MANY SUMMARIES OF DATA, SUCH AS THE SAMPLE MEAN, ARE LINEAR COMBINATIONS OF DATA (I.E., SUMS OF THE TYPE $aX + bY + cZ + ...$)

THE WORLD IS THE SUM OF ITS PARTS!

IN THE NEXT CHAPTER, WE WILL SEE TWO IMPORTANT EXAMPLES OF RANDOM VARIABLES: ONE, THE *BINOMIAL*, IS THE SUM OF MANY REPEATED INDEPENDENT RANDOM VARIABLES. THE OTHER, THE *NORMAL*, IS A CONTINUOUS RANDOM VARIABLE THAT HAS A SURPRISING RELATIONSHIP TO THE BINOMIAL, AND ANY OTHER SUM OF INDEPENDENT RANDOM VARIABLES AS WELL.

JUST REMEMBER: RANDOM EXPERIMENT, NUMERICAL OUTCOME!

MM. SOUNDS LIKE MY LAST PAYCHECK...

◆ Chapter 5 ◆
A TALE OF TWO DISTRIBUTIONS

NOW WE LOOK AT TWO IMPORTANT EXAMPLES OF
RANDOM VARIABLES, ONE DISCRETE AND ONE CONTINUOUS.

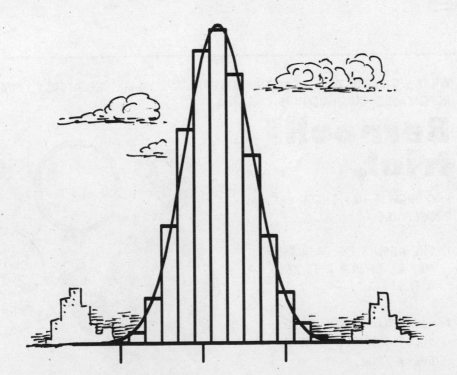

WE BEGIN WITH THE DISCRETE ONE, CALLED THE *BINOMIAL* RANDOM VARIABLE. SUPPOSE WE HAVE A RANDOM PROCESS WITH JUST *TWO POSSIBLE OUTCOMES*: A HEADS-OR-TAILS COIN TOSS, A WIN-OR-LOSE FOOTBALL GAME, A PASS-OR-FAIL AUTOMOTIVE SMOG INSPECTION. WE ARBITRARILY CALL ONE OF THESE OUTCOMES A *SUCCESS* AND THE OTHER A *FAILURE*.

CONGRATULATIONS ON YOUR SUCCESS! YOUR CAR JUST FAILED THE SMOG TEST!

WHAT WE DO IS TO REPEAT THIS EXPERIMENT... WELL, REPEATEDLY. SUCH A REPEATABLE EXPERIMENT IS CALLED A

Bernoulli trial,

PROVIDED IT HAS THESE CRITICAL PROPERTIES:

1) THE RESULT OF EACH TRIAL MAY BE EITHER A SUCCESS OR A FAILURE

2) THE PROBABILITY p OF SUCCESS IS THE SAME IN EVERY TRIAL.

3) THE TRIALS ARE *INDEPENDENT*: THE OUTCOME OF ONE TRIAL HAS NO INFLUENCE ON LATER OUTCOMES.

NO PICTURE OF BERNOULLI... SORRY!

STARTING WITH A BERNOULLI TRIAL, WITH PROBABILITY OF SUCCESS p, LET'S BUILD A NEW RANDOM VARIABLE BY REPEATING THE BERNOULLI TRIAL.

The binomial random variable

X IS THE *NUMBER OF SUCCESSES* IN n REPEATED BERNOULLI TRIALS WITH PROBABILITY p OF SUCCESS.

AN EXAMPLE OF A BINOMIAL RANDOM VARIABLE IS THE NUMBER OF HEADS (SUCCESSES) IN TWO FLIPS OF A COIN. HERE $n = 2$ AND $p = .5$

k = NUMBER OF SUCCESSES	0	1	2
Pr(X=k)	.25	.5	.25

ANOTHER EXAMPLE IS DE MERE'S FIRST GAMBLE: TOSSING A SINGLE DIE FOUR TIMES IN A ROW. SUCCESS MEANS ROLLING A 6. THE DISTRIBUTION IS:

UM... THE DISTRIBUTION IS... IS...?

WHAT IS THE PROBABILITY OF ROLLING k 6'S IN 4 ROLLS?

IN GENERAL, WHAT'S THE PROB-
ABILITY DISTRIBUTION OF THE
BINOMIAL FOR *ANY* PROBABILITY
p AND NUMBER OF TRIALS n? A
PROBABILITY CALCULATION GIVES
THE ANSWER: THE PROBABILITY
OF OBTAINING k SUCCESSES IN
n TRIALS, $Pr(X=k)$, IS

$$Pr(X=k) = \binom{n}{k}p^k(1-p)^{n-k}$$

HERE $\binom{n}{k}$, READ "n CHOOSE k," IS THE **BINOMIAL COEFFICIENT**. IT COUNTS
ALL POSSIBLE WAYS OF GETTING k SUCCESSES IN n TRIALS. EACH INDIVIDUAL
SEQUENCE OF k SUCCESSES AND $n-k$ FAILURES HAS PROBABILITY $p^k(1-p)^{n-k}$,
BY THE MULTIPLICATION RULE. THERE ARE $\binom{n}{k}$ OF THESE SEQUENCES.

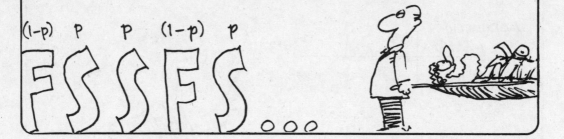

THE FORMULA FOR $\binom{n}{k}$ IS

$$\binom{n}{k} = \frac{n!}{k!(n-k)!}$$

WHERE

$$n! = n \times (n-1) \times (n-2) \times \ldots \times 1$$

AND $0!$ IS TAKEN TO BE 1. FOR INSTANCE,
$\binom{4}{2}$, THE NUMBER OF POSSIBLE WAYS TO
CHOOSE TWO LETTERS FROM A SET OF
FOUR LETTERS, IS

$$\binom{4}{2} = \frac{4!}{2!2!} = \frac{24}{4} = 6$$

76

ANOTHER VIEW OF THE BINOMIAL COEFFICIENTS IS IN *PASCAL'S TRIANGLE.* EACH ENTRY IS THE SUM OF THE TWO NUMBERS JUST ABOVE IT.

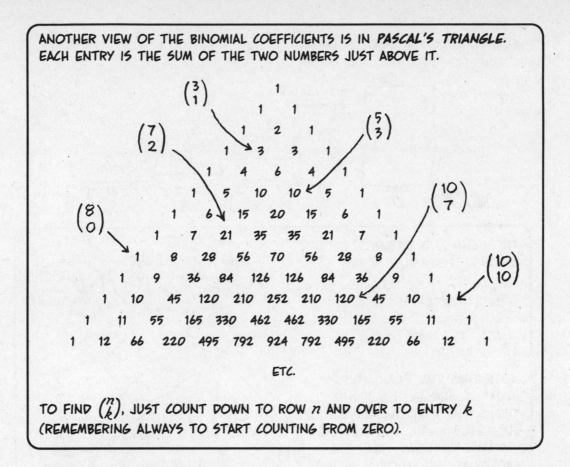

ETC.

TO FIND $\binom{n}{k}$, JUST COUNT DOWN TO ROW n AND OVER TO ENTRY k (REMEMBERING ALWAYS TO START COUNTING FROM ZERO).

WHEN p = .5, THE BINOMIAL'S PROBABILITY DISTRIBUTION IS PERFECTLY SYMMETRICAL. FOR 6 COIN FLIPS, FOR INSTANCE, IT'S

k = #HEADS	0	1	2	3	4	5	6
$Pr(X=k)$	$\left(\frac{1}{2}\right)^6$	$\left(\frac{1}{2}\right)^6 \cdot 6$	$\left(\frac{1}{2}\right)^6 \cdot 15$	$\left(\frac{1}{2}\right)^6 \cdot 20$	$\left(\frac{1}{2}\right)^6 \cdot 15$	$\left(\frac{1}{2}\right)^6 \cdot 6$	$\left(\frac{1}{2}\right)^6$

WITH THIS HISTOGRAM:

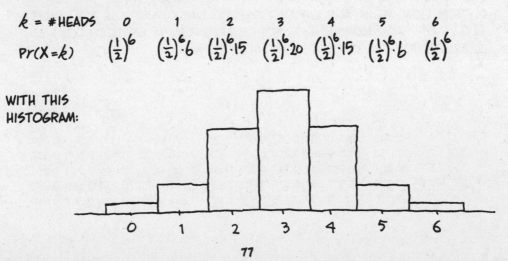

FOR DE MERE'S ROLL OF FOUR DICE, THE DISTRIBUTION IS MORE LOPSIDED:

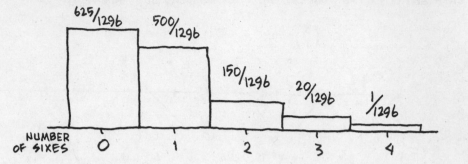

$\frac{625}{1296}$ $\frac{500}{1296}$ $\frac{150}{1296}$ $\frac{20}{1296}$ $\frac{1}{1296}$

NUMBER OF SIXES   0   1   2   3   4

THE **MEAN** AND **VARIANCE** OF THE BINOMIAL DISTRIBUTION ARE

$$\mu = np$$
$$\sigma^2 = np(1-p)$$

NOTE THAT THE MEAN MAKES INTUITIVE SENSE: IN n BERNOULLI TRIALS, THE EXPECTED NUMBER OF SUCCESSES SHOULD BE np. THE VARIANCE FOLLOWS FROM THE FACT THAT THE BINOMIAL IS THE SUM OF n INDEPENDENT BERNOULLI TRIALS OF VARIANCE $p(1-p)$.

THE PARAMETERS OF THE BINOMIAL DISTRIBUTION ARE n AND p. THE DISTRIBUTION, MEAN, AND VARIANCE DEPEND **ONLY** ON THESE TWO NUMBERS. TABLES OF THE BINOMIAL DISTRIBUTION APPEAR IN MOST TEXTBOOKS AND COMPUTER PROGRAMS. HERE IS A TABLE FOR $n=10$.

VALUES OF $Pr(X=k)$

		k									
		0	1	2	3	4	5	6	7	8	9	10
	.1	0.349	0.387	0.194	0.057	0.011	0.001	0.000	0.000	0.000	0.000	0.000
	.25	0.056	0.188	0.282	0.250	0.146	0.058	0.016	0.003	0.000	0.000	0.000
p	.50	0.001	0.010	0.044	0.117	0.205	0.246	0.205	0.117	0.044	0.010	0.001
	.75	0.000	0.000	0.000	0.003	0.016	0.058	0.146	0.250	0.282	0.188	0.056
	.9	0.000	0.000	0.000	0.000	0.000	0.001	0.011	0.057	0.194	0.387	0.349

BUT CALCULATING THESE THINGS FOR LARGE VALUES OF n CAN BE A PAIN... OR AT LEAST, IT WAS BACK IN THE 18TH CENTURY, WHEN *JAMES BERNOULLI* AND *ABRAHAM DE MOIVRE* WERE TRYING TO DO IT WITHOUT A COMPUTER.

WE NEED NEW TOOLS!

OR WIDER PAPER...

DEPLOYING A NEWLY INVENTED WEAPON, THE *CALCULUS*, DE MOIVRE SHOWED THAN WHEN $p = .5$, THE BINOMIAL DISTRIBUTION WAS CLOSELY APPROXIMATED BY A *CONTINUOUS DENSITY FUNCTION* WHICH COULD BE DESCRIBED VERY SIMPLY.

TO SEE HOW THIS WORKS, IMAGINE THE BINOMIAL DISTRIBUTION WITH $p = .5$ AND n VERY LARGE—A MILLION, SAY...

UM... WHAT A WIDE, LOW THING...

NOW, SAID DEMOIVRE, SLIDE THIS GRAPH OVER, SO ITS MEAN IS ZERO.

OR SLIDE THE y-AXIS— SAME THING!

SQUASH THE CURVE ALONG THE x AXIS UNTIL THE STANDARD DEVIATION BECOMES 1, WHILE STRETCHING IT ALONG THE y AXIS TO KEEP THE AREA UNDER IT EQUAL TO 1.

SQUISH

THE RESULT IS VERY CLOSE TO A *SMOOTH, SYMMETRICAL, BELL-SHAPED CURVE,* WHICH DEMOIVRE SHOWED WAS GIVEN BY THE SIMPLE FORMULA:

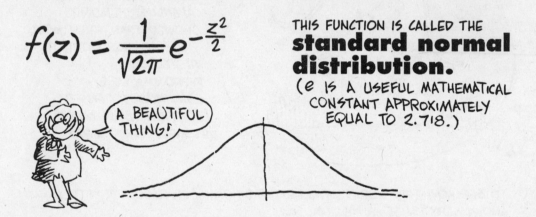

$$f(z) = \frac{1}{\sqrt{2\pi}} e^{-\frac{z^2}{2}}$$

A BEAUTIFUL THING!

THIS FUNCTION IS CALLED THE **standard normal distribution.**

(e IS A USEFUL MATHEMATICAL CONSTANT APPROXIMATELY EQUAL TO 2.718.)

(CONVINCE YOURSELF THAT THIS FUNCTION REALLY HAS A BELL-SHAPED GRAPH. FOR z FAR FROM ZERO, $f(z)$ IS VERY NEARLY ZERO—IT HAS A BIG DENOMINATOR; IT'S SYMMETRICAL, SINCE $f(z) = f(-z)$, AND IT HAS A MAXIMUM AT $z = 0$.)

THE DISTRIBUTION IS CALLED THE *STANDARD* NORMAL BECAUSE ALL THAT SQUASHING AND STRETCHING WAS SPECIALLY ARRANGED TO GIVE IT THESE SIMPLE PROPERTIES, WHICH WE PRESENT WITHOUT PROOF:

$$\mu = 0$$
$$\sigma = 1$$

TO SUMMARIZE DE MOIVRE, IF YOU *"NORMALIZE" THE BINOMIAL* DISTRIBUTION WITH $p = 1/2$—I.E., CENTER IT ON ZERO AND MAKE ITS STANDARD DEVIATION = 1, THEN IT CLOSELY FITS THE *STANDARD NORMAL DISTRIBUTION*

$$f(z) = \frac{1}{\sqrt{2\pi}} e^{-\frac{z^2}{2}}$$

B·BUT... WHAT ABOUT THE C·C·C· CALCULUS??

THAT WAS FOR DEMOIVRE, NOT FOR US...

OTHER NORMALS, WITH DIFFERENT MEANS AND VARIANCES, ARE OBTAINED BY STRETCHING AND SLIDING THE STANDARD NORMAL. IN GENERAL, WE WRITE THE FORMULA

$$f(x \mid \mu, \sigma) = \frac{1}{\sigma\sqrt{2\pi}} e^{-\frac{1}{2}\left(\frac{x-\mu}{\sigma}\right)^2}$$

THIS GIVES A SYMMETRIC, BELL-SHAPED DISTRIBUTION CENTERED ON THE MEAN μ WITH THE STANDARD DEVIATION σ.

HERE ARE TWO DIFFERENT NORMALS WITH THE REGIONS WITHIN THEIR STANDARD DEVIATIONS SHADED.

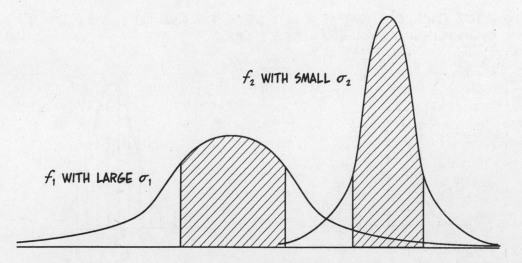

f_2 WITH SMALL σ_2

f_1 WITH LARGE σ_1

DE MOIVRE PROVED THAT THE STANDARD NORMAL FITS THE (NORMALIZED) BINOMIAL WITH $p = .5$, BUT, IN FACT, IT WORKS FOR *ANY VALUE OF* p.

GENERALLY: FOR ANY VALUE OF p, THE BINOMIAL DISTRIBUTION OF n TRIALS WITH PROBABILITY p IS APPROXIMATED BY THE NORMAL CURVE WITH $\mu = np$ AND $\sigma = np(1 - p)$.

OO! OO!

ALL BINOMIALS TURN INTO NORMALS, EVENTUALLY...

A BELL APPROXIMATES THIS?

THIS IS ACTUALLY A LITTLE STRANGE. ALL NORMALS ARE SYMMETRICAL AND BELL SHAPED... BUT, AS WE SAW, BINOMIAL DISTRIBUTIONS ARE *NOT SYMMETRICAL* WHEN $p \neq .5$.

BUT IT TURNS OUT THAT AS n GETS LARGE, THE BINOMIAL'S ASYMMETRY IS OVERWHELMED, AS YOU SEE IN THIS EXAMPLE:

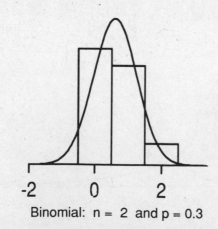

Binomial: n = 2 and p = 0.3

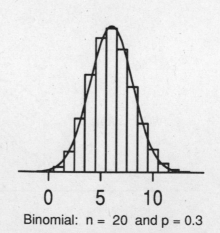

Binomial: n = 20 and p = 0.3

IN FACT, DEMOIVRE'S DISCOVERY ABOUT THE BINOMIAL IS A SPECIAL CASE OF AN EVEN *MORE* GENERAL RESULT, WHICH HELPS EXPLAIN WHY THE NORMAL IS SO IMPORTANT AND WIDESPREAD IN NATURE. IT IS THIS:

"Fuzzy Central Limit Theorem":

DATA THAT ARE INFLUENCED BY *MANY SMALL AND UNRELATED RANDOM EFFECTS* ARE APPROXIMATELY *NORMALLY DISTRIBUTED.*

> MON DIEU! THIS INCLUDES EVERYTHING!!

THIS EXPLAINS WHY THE NORMAL IS *EVERYWHERE*: STOCK MARKET FLUCTUATIONS, STUDENT WEIGHTS, YEARLY TEMPERATURE AVERAGES, S.A.T. SCORES: ALL ARE THE RESULT OF MANY DIFFERENT EFFECTS. FOR EXAMPLE, A STUDENT'S WEIGHT IS THE RESULT OF GENETICS, NUTRITION, ILLNESS, AND LAST NIGHT'S BEER PARTY. WHEN YOU PUT THEM ALL TOGETHER, YOU GET THE NORMAL! (REMEMBER, THE BINOMIAL IS THE RESULT OF n INDEPENDENT BERNOULLI TRIALS.)

> YOU MEAN THIS IS *NORMAL?*

> OORG...NEXT TIME REMIND ME TO STOP AFTER $n-1$ BEERS...

> I FEEL MOUND-SHAPED.

NOW, BACK TO THE MATH....

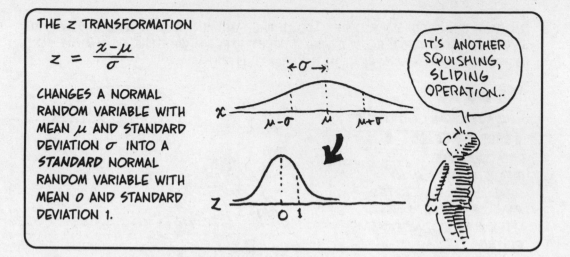

THE z TRANSFORMATION

$$z = \frac{x - \mu}{\sigma}$$

CHANGES A NORMAL RANDOM VARIABLE WITH MEAN μ AND STANDARD DEVIATION σ INTO A *STANDARD* NORMAL RANDOM VARIABLE WITH MEAN 0 AND STANDARD DEVIATION 1.

IT'S ANOTHER SQUISHING, SLIDING OPERATION..

THEN ALL WE NEED TO FIND PROBABILITIES FOR *ANY* NORMAL DISTRIBUTION IS THE SINGLE TABLE FOR THE *STANDARD NORMAL* $f(z)$.

z	-2.5	-2.4	-2.3	-2.2	-2.1	-2.0	-1.9	-1.8	-1.7	-1.6
F(z)	0.006	0.008	0.011	0.014	0.018	0.023	0.029	0.036	0.045	0.055
z	-1.5	-1.4	-1.3	-1.2	-1.1	-1.0	-0.9	-0.8	-0.7	-0.6
F(z)	0.067	0.081	0.097	0.115	0.136	0.159	0.184	0.212	0.242	0.274
z	-0.5	-0.4	-0.3	-0.2	-0.1	0.0	0.1	0.2	0.3	0.4
F(z)	0.309	0.345	0.382	0.421	0.460	0.500	0.540	0.579	0.618	0.655
z	0.5	0.6	0.7	0.8	0.9	1.0	1.1	1.2	1.3	1.4
F(z)	0.691	0.726	0.758	0.788	0.816	0.841	0.864	0.885	0.903	0.919
z	1.5	1.6	1.7	1.8	1.9	2.0	2.1	2.2	2.3	2.4
F(z)	0.933	0.945	0.955	0.964	0.971	0.977	0.982	0.986	0.989	0.992
z	2.5									
F(z)	0.994									

WOOF

HERE $F(a) = Pr(z \leq a)$, THE AREA UNDER THE DENSITY CURVE TO THE LEFT OF $z = a$.

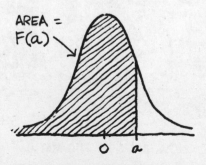

AREA = $F(a)$

(WE CAN ALSO GRAPH THE CURVE $y = F(z)$, THE *CUMULATIVE PROBABILITY*. IT LOOKS LIKE THIS.)

THE TABLE ALLOWS US TO FIND THE PROBABILITY OF z BEING IN ANY INTERVAL $a \leq z \leq b$. IT IS JUST THE DIFFERENCE BETWEEN THE AREAS $F(b)$ AND $F(a)$.

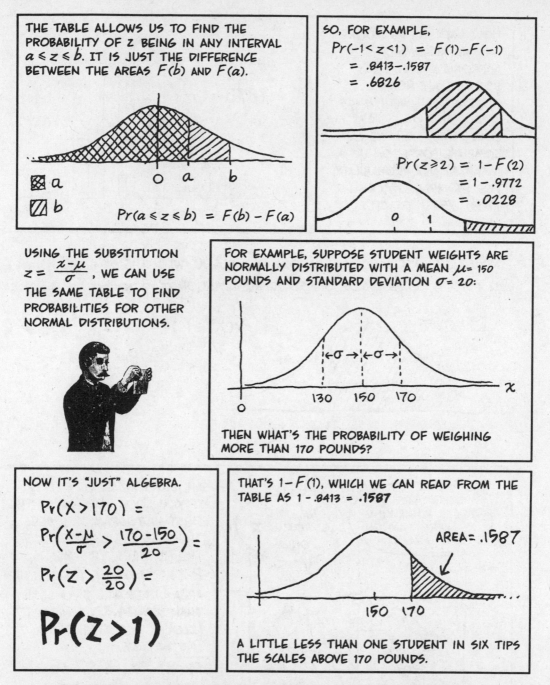

▨ a

▧ b

$$Pr(a \leq z \leq b) = F(b) - F(a)$$

SO, FOR EXAMPLE,

$Pr(-1 < z < 1) = F(1) - F(-1)$
$= .8413 - .1587$
$= .6826$

$Pr(z \geq 2) = 1 - F(2)$
$= 1 - .9772$
$= .0228$

USING THE SUBSTITUTION $z = \frac{x - \mu}{\sigma}$, WE CAN USE THE SAME TABLE TO FIND PROBABILITIES FOR OTHER NORMAL DISTRIBUTIONS.

FOR EXAMPLE, SUPPOSE STUDENT WEIGHTS ARE NORMALLY DISTRIBUTED WITH A MEAN $\mu = 150$ POUNDS AND STANDARD DEVIATION $\sigma = 20$:

THEN WHAT'S THE PROBABILITY OF WEIGHING MORE THAN 170 POUNDS?

NOW IT'S "JUST" ALGEBRA.

$Pr(X > 170) =$

$Pr\left(\frac{X - \mu}{\sigma} > \frac{170 - 150}{20}\right) =$

$Pr\left(Z > \frac{20}{20}\right) =$

$$Pr(Z > 1)$$

THAT'S $1 - F(1)$, WHICH WE CAN READ FROM THE TABLE AS $1 - .8413 = .1587$

AREA = .1587

A LITTLE LESS THAN ONE STUDENT IN SIX TIPS THE SCALES ABOVE 170 POUNDS.

THE GENERAL RULE FOR COMPUTING NORMAL PROBABILITIES IS THEREFORE:

$$Pr(a \leq X \leq b) = F\left(\frac{b - \mu}{\sigma}\right) - F\left(\frac{a - \mu}{\sigma}\right)$$

NOW BACK TO DE MOIVRE AND HIS BINOMIAL APPROXIMATION... LET'S LOOK AT A BINOMIAL DISTRIBUTION WITH $n = 25$ TRIALS AND $p = .5$ (25 COIN FLIPS, SAY). WE CAN COMPUTE (OR LOOK UP IN A TABLE) ANY PROBABILITY, FOR EXAMPLE, $Pr(X \leqslant 14)$. IT IS **.7878** EXACTLY.

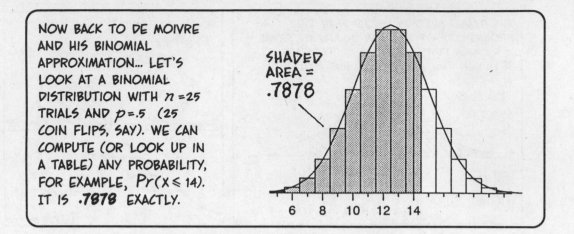

SHADED AREA = .7878

NOW CALCULATE A NORMAL RANDOM VARIABLE X^* WITH THE SAME MEAN $\mu = np = (25)(.5) = 12.5$ AND STANDARD DEVIATION $\sigma = np(1-p) = 2.5$.

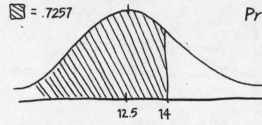

▨ = .7257

$$Pr(X^* \leqslant 14) = Pr(Z \leqslant \frac{14 - 12.5}{2.5})$$
$$= Pr(Z \leqslant .6)$$
$$= \mathbf{.7257}$$

.7878 VERSUS .7257? WHAT KIND OF APPROXIMATION IS *THAT*?

UM... AN *APPROXIMATE* ONE?

AH, BUT WE CAN DO BETTER! IF YOU LOOK CLOSELY AT THE FIRST HISTOGRAM, YOU SEE THE BARS ARE *CENTERED* ON THE NUMBERS. THIS MEANS $Pr(X^* \leqslant 14)$ IS ACTUALLY THE AREA UNDER THE BARS LESS THAN $x = $ **14.5**. WE NEED TO ACCOUNT FOR THAT EXTRA .5, AND IN FACT,

$$Pr(X^* \leqslant 14.5) = Pr(Z \leqslant .8)$$
$$= \mathbf{.7881}$$

A VERY GOOD APPROXIMATION TO .7878 INDEED!

THAT LITTLE EXTRA .5 WE ADDED IS CALLED THE

continuity correction.

WE HAVE TO INCLUDE IT TO GET A GOOD CONTINUOUS APPROXIMATION TO OUR DISCRETE BINOMIAL RANDOM VARIABLE X. IT'S SUMMARIZED BY THIS ONE HIDEOUS FORMULA:

WE HAVE TO GO TO THE EDGES!

$$Pr(a \leq X \leq b) \simeq Pr\left(\frac{a - \frac{1}{2} - np}{\sqrt{np(1-p)}} \leq Z \leq \frac{b + \frac{1}{2} - np}{\sqrt{np(1-p)}}\right)$$

WHEN IS THIS APPROXIMATION "GOOD ENOUGH?" FOR STATISTICIANS, THE RULE OF THUMB IS: WHENEVER n IS BIG ENOUGH TO MAKE THE NUMBER OF EXPECTED SUCCESSES AND FAILURES BOTH GREATER THAN *FIVE*:

$$np \geq 5 \quad and \quad n(1-p) \geq 5$$

YOU CAN SEE FROM THESE HISTOGRAMS THAT THE FIT WHEN $p = 0.1$ IS MEDIOCRE OR WORSE UNTIL n REACHES 50, MAKING $np = 5$.

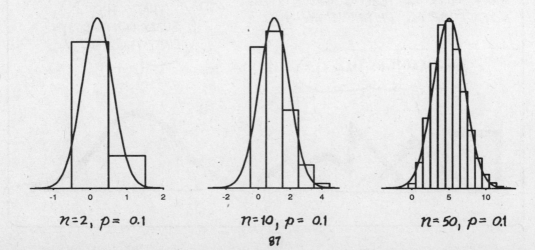

$n = 2, \ p = 0.1$ $n = 10, \ p = 0.1$ $n = 50, \ p = 0.1$

WHAT'S SO GREAT ABOUT THIS NORMAL APPROXIMATION? THE BINOMIAL
DISTRIBUTION OCCURS COMMONLY IN NATURE, AND IT ISN'T HARD TO UNDER-
STAND, BUT IT CAN BE TIRESOME TO CALCULATE.

THERE'S A NEW ONE
FOR EVERY VALUE
OF n AND p...

THE NORMAL WHICH APPROXIMATES IT MAY BE LESS INTUITIVE, BUT IT'S VERY
EASY TO USE. THE z-TRANSFORM CONVERTS *ANY* NORMAL TO THE *STANDARD
NORMAL*, ALLOWING US TO READ PROBABILITIES STRAIGHT OUT OF A SINGLE
NUMERICAL TABLE.

IN A BOOK
OR ON A
COMPUTER
SCREEN!

RECIPES FOR STANDARD NORMALS

OTHER FILLING MEALS

AND BESIDES, THE NORMAL REALLY IS *THE
MOTHER OF ALL DISTRIBUTIONS!*

MOMMY! MOMMY!

THAT'S THE
FUZZY CENTRAL
LIMIT THEOREM!

♦Chapter 6♦
SAMPLING

BY NOW, AFTER A STEADY DIET OF COINS, DICE, AND ABSTRACT IDEAS, YOU MAY BE WONDERING WHAT ALL THIS STATISTICAL EQUIPMENT WE'VE BEEN BUILDING HAS TO DO WITH THE *REAL WORLD*. WELL, NOW WE'RE FINALLY GOING TO FIND OUT...

IN THIS CHAPTER, WE BEGIN LOOKING AT THE *REAL* BUSINESS OF STATISTICS, WHICH IS, AFTER ALL, TO SAVE PEOPLE *TIME* AND *MONEY*. PEOPLE HATE TO WASTE TIME DOING *UNNECESSARY WORK*, AND ONE THING STATISTICS CAN DO IS TELL US EXACTLY HOW LAZY WE CAN AFFORD TO BE.

THE PROBLEM WITH THE WORLD IS THAT THE COLLECTIONS OF STUFF IN IT ARE SO LARGE, IT'S HARD TO GET THE INFORMATION WE WANT:

VOTING POPULATIONS: WHAT PERCENTAGE FAVORS EACH CANDIDATE?

MANUFACTURED GOODS: WHAT PROPORTION WILL BE DEFECTIVE?

PICKLES: WHAT'S THEIR AVERAGE LENGTH?

THE PICKLE-JAR MAKERS NEED TO KNOW!

THE INDUSTRIOUS, HARD-WORKING, SIMPLE-MINDED BEAVERLIKE WAY TO ANSWER THESE QUESTIONS WOULD BE TO MEASURE EVERY SINGLE PICKLE IN THE WORLD (SAY) AND DO SOME ARITHMETIC.

BUT WE AREN'T BEAVERS—WE'RE *STATISTICIANS!* WE'RE LOOKING FOR THE *EASY* WAY OUT...

OH, WELL... I ATE THE PENCIL, ANYWAY...

OUR METHOD IS TO TAKE A *SAMPLE*... A RELATIVELY SMALL SUBSET OF THE TOTAL POPULATION, THE WAY POLLSTERS DO AT ELECTION TIME.

QUESTION ONE: HOW DO YOU FEEL ABOUT POLLING?

AN OBVIOUS QUESTION IS: HOW BIG A SAMPLE DO WE HAVE TO TAKE TO GET MEANINGFUL RESULTS?

GREATER THAN ONE, PROBABLY...

AND THE ANSWER, WHICH YOU SHOULD INSCRIBE IN YOUR BRAIN FOREVERMORE, WILL TURN OUT TO BE: IF n IS THE NUMBER OF ITEMS IN THE SAMPLE, THEN EVERYTHING IS GOVERNED BY

$$\frac{1}{\sqrt{n}}.$$

GOVERNED BY $\frac{1}{\sqrt{n}}$? DIDN'T EVEN KNOW IT WAS ON THE BALLOT!

SAMPLING DESIGN

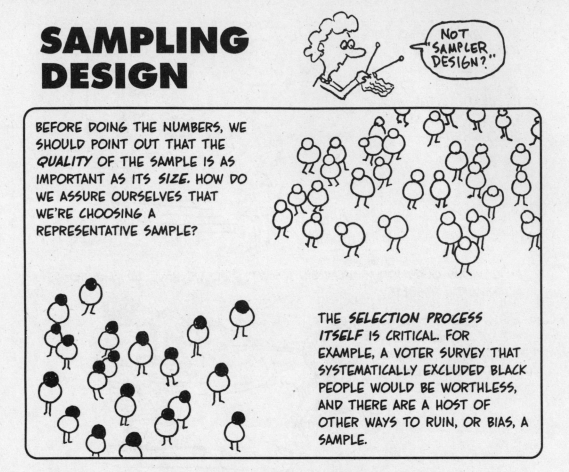

NOT "SAMPLER DESIGN?"

BEFORE DOING THE NUMBERS, WE SHOULD POINT OUT THAT THE *QUALITY* OF THE SAMPLE IS AS IMPORTANT AS ITS *SIZE*. HOW DO WE ASSURE OURSELVES THAT WE'RE CHOOSING A REPRESENTATIVE SAMPLE?

THE *SELECTION PROCESS ITSELF* IS CRITICAL. FOR EXAMPLE, A VOTER SURVEY THAT SYSTEMATICALLY EXCLUDED BLACK PEOPLE WOULD BE WORTHLESS, AND THERE ARE A HOST OF OTHER WAYS TO RUIN, OR BIAS, A SAMPLE.

NOT TO PROLONG THE MYSTERY, THE WAY TO GET STATISTICALLY DEPENDABLE RESULTS IS TO CHOOSE THE SAMPLE AT **random.**

YOU YOU YOU YOU

I CAN'T HEAR YOU! IS IT STILL RANDOM?

THE **SIMPLE** RANDOM SAMPLE

SUPPOSE WE HAVE A LARGE POPULATION OF OBJECTS AND A PROCEDURE FOR SELECTING n OF THEM. IF THE PROCEDURE ENSURES THAT *ALL POSSIBLE SAMPLES* OF n OBJECTS ARE *EQUALLY LIKELY*, THEN WE CALL THE PROCEDURE A **simple random sample.**

GLEEP!

THE SIMPLE RANDOM SAMPLE HAS TWO PROPERTIES THAT MAKE IT THE STANDARD AGAINST WHICH WE MEASURE ALL OTHER METHODS:

RESULTS OF BIASED SAMPLE

1) UNBIASED: EACH UNIT HAS THE SAME CHANCE OF BEING CHOSEN.

2) INDEPENDENCE: SELECTION OF ONE UNIT HAS NO INFLUENCE ON THE SELECTION OF OTHER UNITS.

UNFORTUNATELY, IN THE REAL WORLD, COMPLETELY UNBIASED, INDEPENDENT SAMPLES ARE HARD TO FIND. FOR INSTANCE, SURVEYING VOTERS BY RANDOMLY DIALING TELEPHONE NUMBERS IS BIASED: IT IGNORES VOTERS WITHOUT A TELEPHONE AND OVERSAMPLES PEOPLE WITH MORE THAN ONE NUMBER.

RING

RING

HELLO? PEROT FOR PRESIDENT HEADQUARTERS!

IT'S THEORETICALLY POSSIBLE TO GET A RANDOM SAMPLE BY BUILDING A **SAMPLING FRAME:** A LIST OF EVERY UNIT IN THE POPULATION. BY USING A RANDOM NUMBER GENERATOR, WE CAN PICK n OBJECTS AT RANDOM.

EQUIVALENTLY, WE CAN PUT ALL THE NAMES ON CARDS AND PULL n OF THEM OUT OF A DRUM.

BUT THIS IS NOT ALWAYS EASY. MAKING THE FRAME MAY BE PROHIBITIVELY COSTLY, CONTROVERSIAL, OR EVEN IMPOSSIBLE. FOR EXAMPLE, AN E.P.A. WATER QUALITY STUDY NEEDED A SAMPLING FRAME OF LAKES IN THE U.S., SO THEN SOMEBODY HAS TO DECIDE:

WHAT WET SPOT IS A LAKE?

ARE THERE OTHER WAYS TO SAMPLE THAT ARE MORE *EFFICIENT* AND *COST-EFFECTIVE* THAN A SIMPLE RANDOM SAMPLE? YES—IF YOU ALREADY KNOW SOMETHING ABOUT THE POPULATION. FOR INSTANCE...

Stratified

SAMPLING: DIVIDE THE POPULATION UNITS INTO HOMOGENEOUS GROUPS (STRATA) AND DRAW A SIMPLE RANDOM SAMPLE FROM EACH GROUP.

FOR EXAMPLE, THE POPULATION OF ALL *PICKLES* CAN BE STRATIFIED BY *TYPE OF PICKLE*. WITHIN EACH TYPE OR STRATUM, THE SIZE SHOULD BE LESS VARIABLE.

Cluster
SAMPLING GROUPS THE POPULATION INTO SMALL CLUSTERS, DRAWS A SIMPLE RANDOM SAMPLE OF CLUSTERS, AND OBSERVES EVERYTHING IN THE SAMPLED CLUSTERS. THIS CAN BE COST-EFFECTIVE IF TRAVEL COSTS BETWEEN RANDOMLY SAMPLED UNITS IS HIGH.

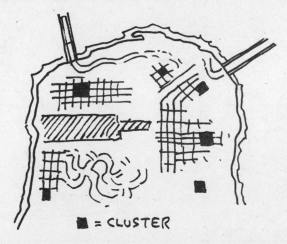

■ = CLUSTER

AN EXAMPLE IS A CITY HOUSING SURVEY WHICH DIVIDES A CITY INTO BLOCKS, RANDOMLY SAMPLES THE BLOCKS, AND LOOKS AT EVERY HOUSING UNIT IN EACH SAMPLED BLOCK.

Systematic SAMPLING STARTS WITH A RANDOMLY CHOSEN UNIT AND THEN SELECTS EVERY k^{TH} UNIT THEREAFTER. FOR INSTANCE, A *HIGHWAY TRAFFIC STUDY* MIGHT CHECK EVERY HUNDREDTH CAR AT A TOLL BOOTH. THIS PLAN IS EASY TO IMPLEMENT AND CAN BE MORE EFFICIENT IF TRAFFIC PATTERNS VARY SMOOTHLY OVER TIME.

EXCUSE ME... WOULD YOU MIND ANSWERING FIFTY OR SIXTY QUESTIONS?

Word of warning #1:

MOST STATISTICAL METHODS DEPEND ON THE INDEPENDENCE AND LACK OF BIAS OF THE SIMPLE RANDOM SAMPLE. THE RESULTS AHEAD APPLY TO THE SIMPLE RANDOM SAMPLE *ONLY*. FOR OTHER SAMPLING PROCEDURES, THE RESULTS MUST BE MODIFIED. THE DETAILS APPEAR IN SPECIALIZED SAMPLING TEXTBOOKS AND COMPUTER ALGORITHMS.

Word of warning #2:

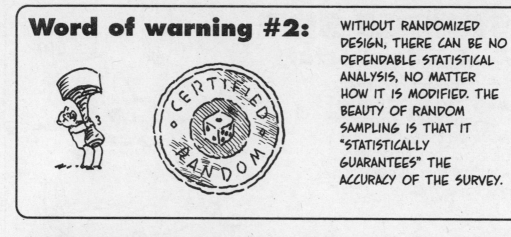

WITHOUT RANDOMIZED DESIGN, THERE CAN BE NO DEPENDABLE STATISTICAL ANALYSIS, NO MATTER HOW IT IS MODIFIED. THE BEAUTY OF RANDOM SAMPLING IS THAT IT "STATISTICALLY GUARANTEES" THE ACCURACY OF THE SURVEY.

A COMMONLY USED METHOD IS ESPECIALLY PRONE TO BIAS: IT'S CALLED AN **opportunity** SAMPLE. AVOIDING ALL THE BOTHER OF DESIGNING A PROCEDURE, THE OPPORTUNITY SAMPLER JUST GRABS THE FIRST n POPULATION UNITS TO COME ALONG.

DON'T WORRY! WE VOLUNTEERED!

A CLASSIC EXAMPLE IS SHERE HITE'S BOOK, *WOMEN AND LOVE*. 100,000 QUESTIONNAIRES WENT TO WOMEN'S ORGANIZATIONS (AN *OPPORTUNITY SAMPLE*), ONLY *4.5%* WERE FILLED OUT AND RETURNED (*RESPONSE BIAS*). SO HER "RESULTS" WERE BASED ON A SAMPLE OF WOMEN WHO WERE HIGHLY MOTIVATED TO ANSWER THE SURVEY'S QUESTIONS, FOR WHATEVER REASON.

AT LAST, A SCIENTIFIC WAY TO HUMILIATE ARNOLD!

SAMPLE SIZE
& standard error

NOW LET'S GET DOWN TO
BRASS TACKS... REAL BRASS
TACKS, THAT IS. SUPPOSE THE
BERNOULLI TACK FACTORY IS
CHURNING OUT BRASS TACKS,
SOME OF WHICH, INEVITABLY,
ARE *DEFECTIVE*.

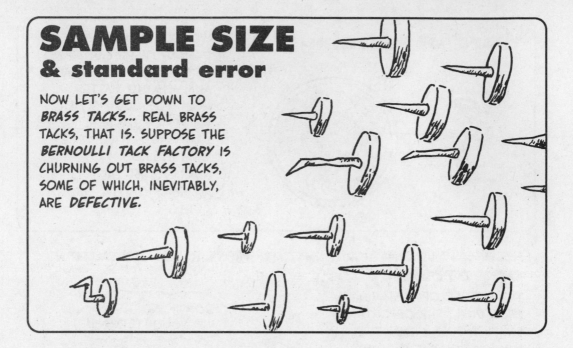

THE ASTUTE READER WILL RECOGNIZE THIS AS A *BERNOULLI SYSTEM*: EACH
NEW TACK IS THE OUTCOME OF A BERNOULLI TRIAL WITH SOME PROBABILITY p
OF SUCCESS (I.E., BEING DEFECT-FREE) AND PROBABILITY $1-p$ OF FAILURE
(I.E., BEING DEFECTIVE).

WE THINK OF THIS SITUATION AS IF THERE WERE A *HIDDEN BUT REAL*
"BERNOULLI MACHINE" WHOSE PROBABILITY p GOVERNS THE OUTCOMES WE
OBSERVE IN THE SO-CALLED "REAL WORLD."

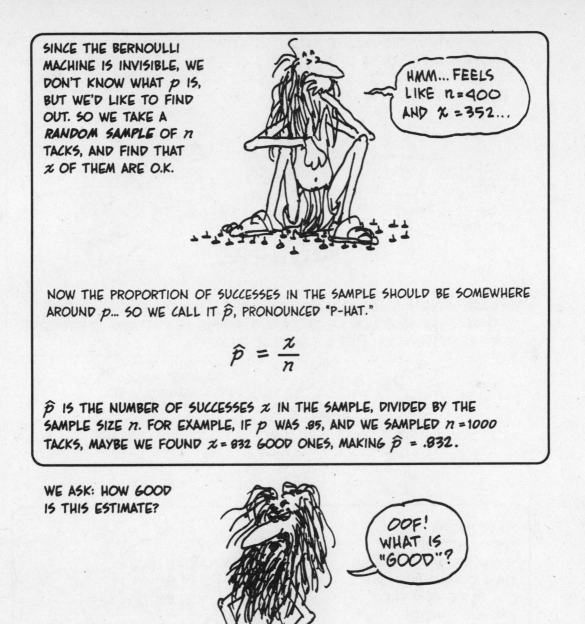

SINCE THE BERNOULLI MACHINE IS INVISIBLE, WE DON'T KNOW WHAT p IS, BUT WE'D LIKE TO FIND OUT. SO WE TAKE A *RANDOM SAMPLE* OF n TACKS, AND FIND THAT x OF THEM ARE O.K.

HMM... FEELS LIKE $n = 400$ AND $x = 352$...

NOW THE PROPORTION OF SUCCESSES IN THE SAMPLE SHOULD BE SOMEWHERE AROUND p... SO WE CALL IT $\hat{p}$, PRONOUNCED "P-HAT."

$$\hat{p} = \frac{x}{n}$$

$\hat{p}$ IS THE NUMBER OF SUCCESSES x IN THE SAMPLE, DIVIDED BY THE SAMPLE SIZE n. FOR EXAMPLE, IF p WAS .85, AND WE SAMPLED $n = 1000$ TACKS, MAYBE WE FOUND $x = 832$ GOOD ONES, MAKING $\hat{p} = .832$.

WE ASK: HOW GOOD IS THIS ESTIMATE?

OOF! WHAT IS "GOOD"?

AND WE ANSWER WITH ANOTHER QUESTION: WHAT DOES THE FIRST QUESTION MEAN?

WE CAN'T KNOW THE PRECISE DIFFERENCE BETWEEN $\hat{p}$ AND p, BECAUSE WE DON'T KNOW THE VALUE OF p. THE **REAL** QUESTION IS THIS: IF WE TOOK **MANY SAMPLES OF 1000 TACKS** AND OBSERVED $\hat{p}$ FOR EACH SAMPLE, HOW WOULD THOSE VALUES OF $\hat{p}$ BE **DISTRIBUTED** AROUND p?

IN FACT, THESE $\hat{p}$ VALUES ARE LOOKING MORE AND MORE LIKE A **RANDOM VARIABLE:** THE SELECTION OF THE n-UNIT SAMPLE IS A RANDOM EXPERIMENT, AND THE OBSERVATION $\hat{p}$ IS A NUMERICAL OUTCOME!

I AM BECOMING ENLIGHTENED NOW... I KNEW IT WOULD NOT BE PAINLESS...

TO BE PRECISE, IF X IS THE NUMBER OF SUCCESSES IN THE SAMPLE, THEN X IS NOTHING BUT OUR OLD FRIEND THE BINOMIAL RANDOM VARIABLE (n TRIALS, PROBABILITY p)... AND WE DEFINE THE **OBSERVED PROPORTION** TO BE THE RANDOM VARIABLE

$$\hat{P} = \frac{X}{n}$$

BIG $\hat{P}$ THE RANDOM VARIABLE, LITTLE $\hat{p}$, ITS VALUE FOR A PARTICULAR SAMPLE!

KNOWING ALL ABOUT X, WE QUICKLY CONCLUDE A FEW FACTS ABOUT $\hat{P}$:

1) THE MEAN OF $\hat{P}$ IS $E[\hat{P}] = p$

2) THE STANDARD DEVIATION OF $\hat{P}$ IS

$$\sigma(\hat{P}) = \frac{\sqrt{p(1-p)}}{\sqrt{n}}$$

3) FOR LARGE n, $\hat{P}$ IS APPROXIMATELY NORMAL.

DEMOIVRE, YOU'RE A GENIUS!

AND THERE YOU HAVE IT ALL! THE OBSERVED VALUES OF $\hat{P}$ WILL BE CENTERED ON p (NOT SURPRISINGLY), AND THEIR STANDARD DEVIATION, OR SPREAD, IS PROPORTIONAL TO THAT MAGIC NUMBER WE MENTIONED AT THE BEGINNING OF THE CHAPTER:

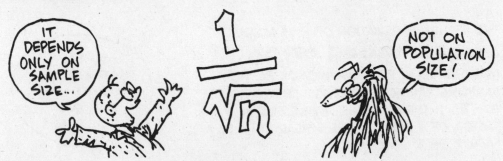

IT DEPENDS ONLY ON SAMPLE SIZE...

$$\frac{1}{\sqrt{n}}$$

NOT ON POPULATION SIZE!

AND, SINCE $\hat{P}$ IS NEARLY NORMAL, WE CAN USE OUR RULE OF THUMB TO CONCLUDE THAT APPROXIMATELY 68% OF ALL ESTIMATES WILL FALL WITHIN ONE STANDARD DEVIATION OF THE TRUE VALUE p.

I'M NEARLY NORMAL, TOO...

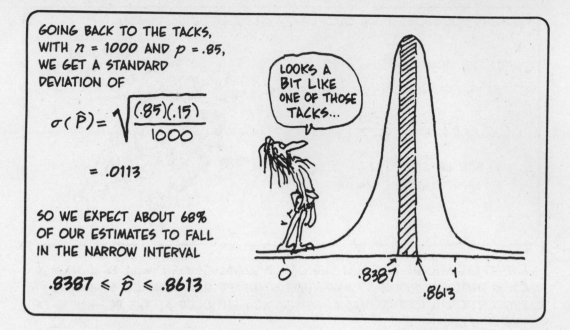

GOING BACK TO THE TACKS, WITH $n = 1000$ AND $p = .85$, WE GET A STANDARD DEVIATION OF

$$\sigma(\hat{P}) = \sqrt{\frac{(.85)(.15)}{1000}}$$

$$= .0113$$

SO WE EXPECT ABOUT 68% OF OUR ESTIMATES TO FALL IN THE NARROW INTERVAL

$$.8387 \leq \hat{p} \leq .8613$$

LOOKS A BIT LIKE ONE OF THOSE TACKS...

THE STANDARD DEVIATION OF $\hat{P}$ IS A MEASURE OF THE **sampling error.** AS WE'VE SEEN, FOR THE BINOMIAL $\hat{P}$, THIS SAMPLING ERROR IS INVERSELY PROPORTIONAL TO $\sqrt{n}$. INCREASING THE SAMPLE SIZE BY A FACTOR OF 4 REDUCES THE SPREAD $\sigma(\hat{P})$ BY A FACTOR OF 2.

ALREADY AT $n = 100$, YOU SEE $\sigma(\hat{p})$ IS DOWN TO $3\frac{1}{2}\%$!

SAMPLE SIZES FOR TACKS, $p = 0.85$

n	1	4	16	25	100	10,000
$\sqrt{n}$	1	2	4	5	10	100
$\sigma(\hat{p})$	.357	.1785	.089	.071	.0357	.0036

LINGUISTIC NOTE: AN *ESTIMATE* IS A SINGLE MEASURE OR OBSERVATION. AN *ESTIMATOR* IS A RULE FOR GETTING ESTIMATES. IN THIS CASE, THE ESTIMATOR IS THE RANDOM VARIABLE $\hat{P} = \frac{X}{n}$.

MOST OF STATISTICS INVOLVES THE 4-STEP PROCESS WE'VE JUST WALKED THROUGH:

DEFINE POPULATION WITH UNKNOWN PARAMETER

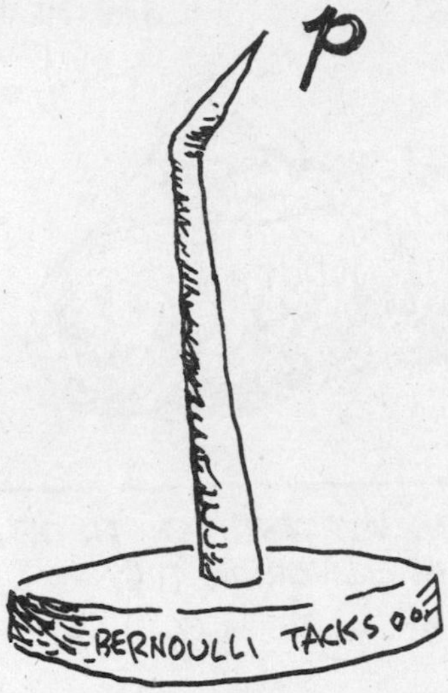

p

BERNOULLI TACKS 00¢

FIND AN ESTIMATOR, ITS THEORETICAL SAMPLING DISTRIBUTION AND STANDARD DEVIATION.

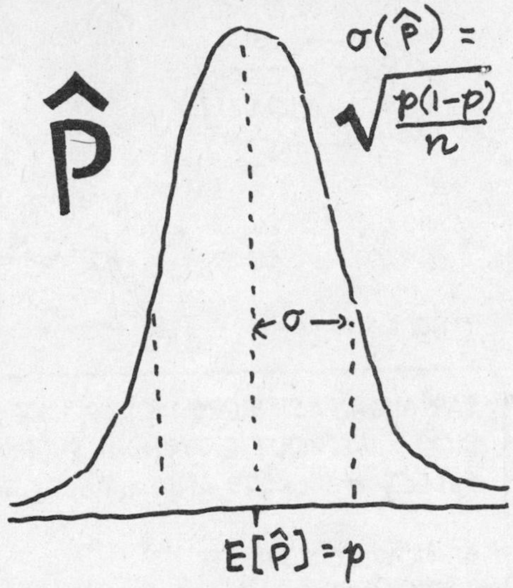

$\hat{p}$

$$\sigma(\hat{p}) = \sqrt{\frac{p(1-p)}{n}}$$

$\leftarrow \sigma \rightarrow$

$E[\hat{p}] = p$

ACTUALLY DRAW A RANDOM SAMPLE AND FIND THE ESTIMATE.

AH... $\hat{p}$ IS EXACTLY .84

REPORT THE RESULT AND ITS STATISTICAL OR SAMPLING ERROR.

WE HAVE $\hat{p}$ = .84 WITH A SAMPLING ERROR OF 1.1%, MR. BERNOULLI, SAHIB...

JUST ONE QUESTION: WHO HIRED YOU?

Sampling Distribution of the MEAN

NOW WE MOVE FROM BRASS TACKS TO DILL PICKLES...

THAT IS ONE MEAN PICKLE!

SLIGHTLY ABOVE THE MEAN, ACTUALLY...

JAR MANUFACTURERS WOULD LIKE TO KNOW THE *AVERAGE LENGTH* OF A PICKLE WITHOUT EXAMINING EVERY CUCUMBER IN CALIFORNIA. THEY RANDOMLY SELECT n PICKLES AND MEASURE THEIR LENGTHS $x_1, x_2, ..., x_n$.

BY NOW YOU MAY BE USED TO THE IDEA THAT EACH X_i IS A *RANDOM VARIABLE:* THE NUMERICAL OUTCOME OF A RANDOM EXPERIMENT.

IF μ IS THE (UNKNOWN) MEAN PICKLE LENGTH, AND σ IS THE STANDARD DEVIATION OF THE *PICKLE LENGTH DISTRIBUTION,* THEN

$$E[X_i] = \mu$$
$$\sigma(X_i) = \sigma$$

FOR EVERY i (BECAUSE x_i COULD HAVE BEEN THE LENGTH OF ANY PICKLE).

STRANGE, HOW MUCH WE KNOW ABOUT RANDOM VARIABLES WE DIDN'T EVEN KNOW WERE RANDOM VARIABLES A MINUTE AGO...

NOW WE LOOK AT THE SAMPLE MEAN: THE AVERAGE LENGTH OF THE SELECTED PICKLES. IT'S A NEW RANDOM VARIABLE GIVEN BY:

$$\overline{X} = \frac{X_1 + X_2 + \ldots + X_n}{n}$$

IS THERE ANYTHING THAT *ISN'T* A RANDOM VARIABLE?

AS BEFORE, WE'D LIKE TO KNOW "HOW CLOSE" THIS IS TO μ, MEANING, IF THIS SAMPLING WERE DONE MANY TIMES, WHAT'S THE DISTRIBUTION OF $\overline{X}$? BECAUSE WE KNOW ABOUT X_1, X_2, ..., AND X_n, WE ALSO KNOW THAT

$$E[\overline{X}] = \mu$$

$$\sigma(\overline{X}) = \sigma/\sqrt{n}$$

THE VARIANCES OF $\frac{X_i}{n}$ ADD TO GIVE THE VARIANCE OF $\overline{X}$

ONCE AGAIN, WE SEE THE MAGIC DENOMINATOR! THE SPREAD OF OBSERVED SAMPLE MEANS GOES AS

$$\frac{1}{\sqrt{n}}.$$

BUT WE DON'T KNOW THE SHAPE OF $\overline{X}$'S DISTRIBUTION. THE SAMPLE *PROBABILITY DISTRIBUTION* $\hat{p}$ WAS ALMOST NORMAL, BECAUSE IT WAS BASED ON A BINOMIAL RANDOM VARIABLE. BUT WHAT ABOUT $\overline{X}$, THE SAMPLE *MEAN* ESTIMATOR???

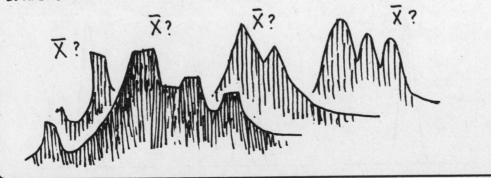

IT TURNS OUT THAT $\bar{X}$ IS ALSO APPROXIMATELY NORMAL! THIS FAMOUS RESULT IS CALLED THE

CENTRAL LIMIT THEOREM

IT SAYS: IF ONE TAKES RANDOM SAMPLES OF SIZE n FROM A POPULATION OF MEAN μ AND STANDARD DEVIATION σ, THEN, AS n GETS LARGE, $\bar{X}$ APPROACHES THE **NORMAL DISTRIBUTION** WITH MEAN μ AND STANDARD DEVIATION $\sigma/\sqrt{n}$. THEN

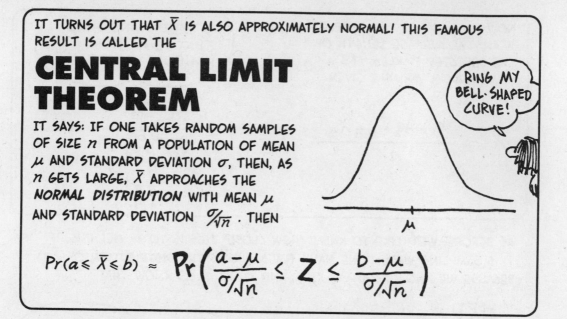

RING MY BELL-SHAPED CURVE!

$$Pr(a \leq \bar{X} \leq b) \approx Pr\left(\frac{a-\mu}{\sigma/\sqrt{n}} \leq Z \leq \frac{b-\mu}{\sigma/\sqrt{n}}\right)$$

WHAT IS REMARKABLE ABOUT THIS? IT SAYS THAT REGARDLESS OF THE SHAPE OF THE ORIGINAL DISTRIBUTION (IN THIS CASE, OF PICKLE LENGTHS), THE TAKING OF **AVERAGES** RESULTS IN A **NORMAL**. TO FIND THE DISTRIBUTION OF $\bar{X}$, WE NEED KNOW **ONLY** THE POPULATION MEAN AND STANDARD DEVIATION.

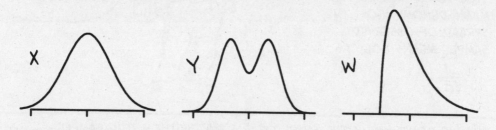

THE THREE PROBABILITY DENSITIES ABOVE ALL HAVE THE SAME MEAN AND STANDARD DEVIATION. DESPITE THEIR DIFFERENT SHAPES, WHEN $n=10$, THE SAMPLING DISTRIBUTIONS OF THE MEAN, $\bar{X}$, ARE NEARLY IDENTICAL.

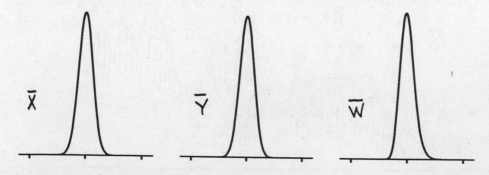

The t-distribution

AMAZING AS THE CENTRAL LIMIT THEOREM IS, IT HAS AT LEAST TWO PROBLEMS.

ONE: IT DEPENDS ON A LARGE SAMPLE SIZE.

TWO: TO USE IT, WE NEED TO KNOW σ, THE STANDARD DEVIATION.

BUT SAMPLE SIZES ARE OFTEN SMALL, AND σ IS USUALLY UNKNOWN. CERTAINLY, IN THE CASE OF THE PICKLES, WE HAVE *NO IDEA* HOW WIDELY THEIR LENGTHS VARY AROUND THE AVERAGE.

WHAT WE CAN DO IN THIS CASE IS TO *ESTIMATE* σ BY TAKING THE *STANDARD DEVIATION OF THE SAMPLE*, WHICH, YOU'LL RECALL, IS GIVEN BY THE FORMULA

$$s = \frac{1}{n-1}\sum_{i=1}^{n}(x_i-\bar{x})^2$$

THEN, IN PLACE OF THE RANDOM VARIABLE

$$z = \frac{\bar{X}-\mu}{\sigma/\sqrt{n}}$$

WE SUBSTITUTE s FOR σ, AND DEFINE A *NEW RANDOM VARIABLE* t BY

$$t = \frac{\bar{X}-\mu}{s/\sqrt{n}}$$

107

YOU CAN THINK OF THE RANDOM VARIABLE t AS *THE BEST WE CAN DO UNDER THE CIRCUMSTANCES.* ITS DISTRIBUTION IS CALLED *STUDENT'S* t, BECAUSE ITS INVENTOR, *WILLIAM GOSSET,* PUBLISHED UNDER THE PSEUDONYM "STUDENT."

GOSSET, YOU IMPLY THAT OUR PRODUCT VARIES IN EXCELLENCE!

PSEUDONYMIZE YOURSELF...

I NEED A P.C... AND A COFFEE...

(GOSSET WAS EMPLOYED BY THE *GUINNESS BREWERY,* WHICH REQUIRED HIM TO USE A PSEUDONYM, FOR SOME REASON.)

MAKING THE ASSUMPTION THAT THE *ORIGINAL POPULATION DISTRIBUTION WAS NORMAL,* OR NEARLY NORMAL, "STUDENT" WAS ABLE TO CONCLUDE:

THE STUFF GETS YOU DRUNK, NO MATTER HOW LOUSY!

t IS MORE SPREAD OUT THAN Z. IT'S "FLATTER" THAN NORMAL. THIS IS BECAUSE THE USE OF S INTRODUCES MORE UNCERTAINTY, MAKING t "SLOPPIER" THAN Z.

Z DIST.

t·DIST

0

THE AMOUNT OF SPREAD DEPENDS ON THE *SAMPLE SIZE.* THE GREATER THE SAMPLE SIZE, THE MORE CONFIDENT WE CAN BE THAT S IS NEAR σ, AND THE CLOSER t GETS TO Z, THE NORMAL.

NORMAL

LARGER SAMPLE t

SMALLER SAMPLE t

0

GOSSET WAS ABLE TO COMPUTE TABLES OF t FOR VARIOUS SAMPLE SIZES, WHICH WE WILL SEE HOW TO USE IN THE FOLLOWING CHAPTER.

IN THE MEANTIME, JUST THINK OF WHAT YOU'VE ALREADY LEARNED!

IN THIS CHAPTER, WE CONSIDERED A CENTRAL PROBLEM OF *REAL-WORLD STATISTICS:* HOW TO SELECT A *SAMPLE* FROM A LARGE POPULATION SO THAT STATISTICAL ANALYSIS CAN BE VALID. BESIDES THE "GOLD STANDARD" OF THE SIMPLE RANDOM SAMPLE, WE ALSO DESCRIBED SOME OTHER SAMPLING SCHEMES THAT ARE USED IN THE INTERESTS OF EFFICIENCY, COST, AND PRACTICALITY.

THEN, ASSUMING A SIMPLE RANDOM SAMPLE, WE CONSIDERED HOW VARIOUS SAMPLE STATISTICS WERE *DISTRIBUTED.* THAT IS, WE REGARDED THE ACT OF TAKING THE SAMPLE AS A *RANDOM EXPERIMENT,* SO THAT ITS STATISTICS BECAME *RANDOM VARIABLES.*

WE FOUND THAT SAMPLE *PROPORTIONS* $\hat{p}$ WERE APPROXIMATELY NORMALLY DISTRIBUTED, WHILE THE DISTRIBUTION OF THE SAMPLE *MEAN* $\bar{X}$ DEPENDED ON THE SAMPLE SIZE. FOR LARGE SAMPLES, THE DISTRIBUTION WAS APPROXIMATELY NORMAL, WHILE FOR SMALL SAMPLES, WE USE THE STUDENT'S t DISTRIBUTION.

IN THE NEXT TWO CHAPTERS, WE LOOK
AT HOW TO USE THESE DISTRIBUTIONS TO
MAKE *STATISTICAL INFERENCES*: GIVEN A
SINGLE OBSERVATION, LIKE A POLITICAL
POLL, HOW DO WE USE OUR KNOWLEDGE
OF $\hat{p}$ AND $\bar{X}$ TO EVALUATE IT?

◆Chapter 7◆
CONFIDENCE INTERVALS

IN THE LAST CHAPTER WE LOOKED AT SAMPLING. STARTING WITH A LARGE POPULATION, WE IMAGINED TAKING MANY SAMPLES, AND WE DEDUCED HOW SOME SAMPLE ESTIMATORS WERE DISTRIBUTED.

$$\sigma = \sqrt{\frac{p(1-p)}{n}}$$

$\hat{p}$

p

IN THIS CHAPTER, WE DO THE REVERSE. GIVEN *ONE* SAMPLE, WE ASK THE QUESTION, WHAT WAS THE RANDOM SYSTEM THAT GENERATED ITS STATISTICS?

THAT IS, GIVEN A SINGLE BOX OF TACKS, AND THE RESULTS OF THE LAST CHAPTER, WHAT CAN WE CONCLUDE?

THIS SHIFT REPRESENTS A CHANGE IN OUR MODE OF THINKING—FROM DEDUCTIVE REASONING TO INDUCTION.

IT'S LIKE A CRIMINAL INVESTIGATION, WATSON!

IN *DEDUCTIVE REASONING*, WE REASON FROM A HYPOTHESIS TO A CONCLUSION: "IF LORD FASTBACK COMMITTED MURDER, THEN HE WOULD WIPE THE FINGERPRINTS OFF THE GUN."

INDUCTIVE REASONING, BY CONTRAST, ARGUES *BACKWARD* FROM A SET OF OBSERVATIONS TO A REASONABLE HYPOTHESIS:

HM. LORD FASTBACK'S *MONOGRAM* ON THIS *HANDKERCHIEF* AND THIS *GUN*. FASTBACK IS THE MURDERER, WATSON, I'M *95% CERTAIN!*

BRILLIANT INDUCTION, HOLMES!

IN MANY WAYS, SCIENCE, INCLUDING STATISTICS, IS LIKE DETECTIVE WORK. BEGINNING WITH A SET OF OBSERVATIONS, WE ASK WHAT CAN BE SAID ABOUT THE SYSTEMS THAT GENERATED THEM.

ESTIMATING
CONFIDENCE INTERVALS

IS ONE OF THE MOST
EFFECTIVE FORMS OF
STATISTICAL INFERENCE,
AND ONE YOU SEE EVERY
DAY BEFORE *ELECTION
TIME*...

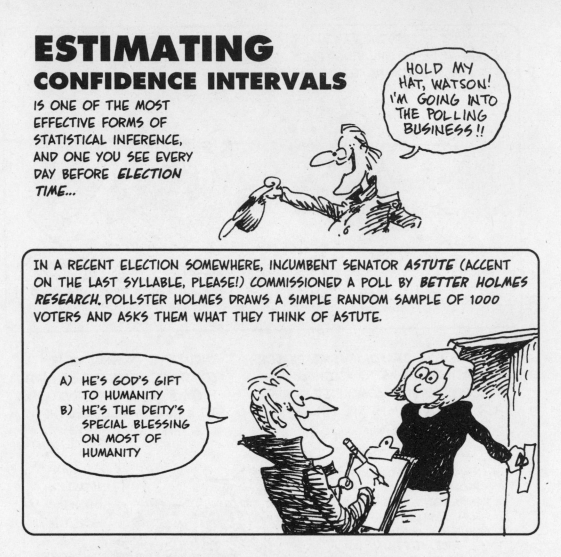

HOLD MY HAT, WATSON! I'M GOING INTO THE POLLING BUSINESS!!

IN A RECENT ELECTION SOMEWHERE, INCUMBENT SENATOR *ASTUTE* (ACCENT ON THE LAST SYLLABLE, PLEASE!) COMMISSIONED A POLL BY *BETTER HOLMES RESEARCH*. POLLSTER HOLMES DRAWS A SIMPLE RANDOM SAMPLE OF 1000 VOTERS AND ASKS THEM WHAT THEY THINK OF ASTUTE.

A) HE'S GOD'S GIFT TO HUMANITY
B) HE'S THE DEITY'S SPECIAL BLESSING ON MOST OF HUMANITY

AFTER CENSORING THE REMARKS OF A FEW GRUMPY OUTLIERS, HOLMES FINDS THAT **550** VOTERS FAVOR HIS CLIENT, SENATOR ASTUTE.

$n = 1000$
$\hat{p} = .55$

THIS IS THE SINGLE OBSERVATION.

AFTER ASTUTE CALMS DOWN, HOLMES EXPLAINS WHAT HE MEANS BY *95% CONFIDENCE*: HE KNOWS THAT HIS ESTIMATION PROCEDURE HAS A *95% PROBABILITY* OF PRODUCING AN INTERVAL CONTAINING p, I.E., IN HIS MANY YEARS OF POLLING, p HAS FALLEN WITHIN THE CONFIDENCE INTERVAL AROUND THE OBSERVED VALUE, $\hat{p}$, 95% OF THE TIME.

SENATOR ASTUTE IS STILL CONFUSED! SO HOLMES GIVES HIM AN **archery lesson.**

SHOOT! ANYTHING TO TAKE MY MIND OFF THEM DANG STATISTICS!

CONSIDER AN ARCHER-POLLSTER SHOOTING AT A TARGET. SUPPOSE THAT SHE HITS THE 10 CM RADIUS BULL'S-EYE 95% OF THE TIME. THAT IS, ONLY ONE ARROW OUT OF 20 MISSES.

SITTING BEHIND THE TARGET IS A BRAVE DETECTIVE, WHO CAN'T SEE THE BULL'S-EYE. THE ARCHER SHOOTS A SINGLE ARROW.

KNOWING THE ARCHER'S SKILL LEVEL, THE DETECTIVE DRAWS A CIRCLE WITH 10 CM RADIUS AROUND THE ARROW. HE NOW HAS *95% CONFIDENCE* THAT HIS CIRCLE INCLUDES THE CENTER OF THE BULL'S-EYE!

10 CM

HE REASONED THAT IF HE DREW 10 CM RADIUS CIRCLES AROUND *MANY* ARROWS, HIS CIRCLES WOULD INCLUDE THE CENTER 95% OF THE TIME.

TRUE CENTER

X = ARROW

(PROBABILISTS USE THE TERM *STOCHASTIC* TO DESCRIBE RANDOM MODELS. IT'S DERIVED FROM THE GREEK *STOCHAZES-THAI,* MEANING TO AIM AT A TARGET, OR GUESS, FROM *STOCHOS,* A TARGET.)

IT'S A TWO-STEP THING!

GO SLOW! I AIN'T USED TO SOMEBODY ELSE LEADING...

HOLMES NOW TRANSLATES THE ARCHERY LESSON INTO THE LANGUAGE WE DEVELOPED LAST CHAPTER.

Step One: SHOOT A LOT OF ARROWS.

A PROBABILITY CALCULATION FINDS THE WIDTH OF THE "BULL'S-EYE." THE ESTIMATES $\hat{p}$ ARE OUR ARROWS. WE SAW THAT THE SAMPLING DISTRIBUTION OF $\hat{p}$ IS NEARLY *NORMAL* WITH MEAN p AND STANDARD DEVIATION

$$\sigma(\hat{p}) = \frac{\sqrt{p(1-p)}}{\sqrt{n}}$$

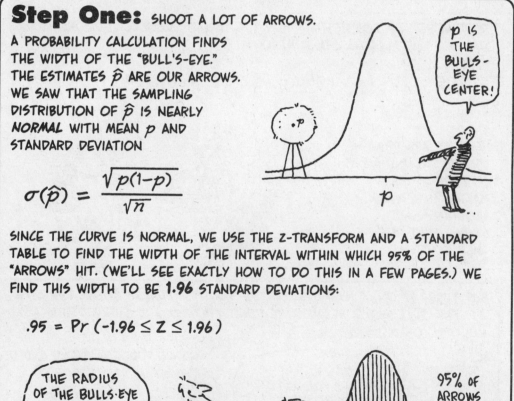

p IS THE BULLS-EYE CENTER!

SINCE THE CURVE IS NORMAL, WE USE THE Z-TRANSFORM AND A STANDARD TABLE TO FIND THE WIDTH OF THE INTERVAL WITHIN WHICH 95% OF THE "ARROWS" HIT. (WE'LL SEE EXACTLY HOW TO DO THIS IN A FEW PAGES.) WE FIND THIS WIDTH TO BE **1.96** STANDARD DEVIATIONS:

$$.95 = \Pr(-1.96 \leq Z \leq 1.96)$$

THE RADIUS OF THE BULLS-EYE IS 1.96 STANDARD DEVIATIONS.

95% OF ARROWS LAND WITHIN THIS INTERVAL

NOW WE DO SOME ALGEBRA. BY DEFINITION OF THE Z-TRANSFORM,

$$.95 \simeq \Pr\left(-1.96 \leq \frac{\hat{p} - p}{\sigma(p)} \leq 1.96\right)$$

WHICH BECOMES

$$.95 \simeq \Pr\left(p - 1.96\,\sigma(p) \leq \hat{p} \leq p + 1.96\,\sigma(p)\right)$$

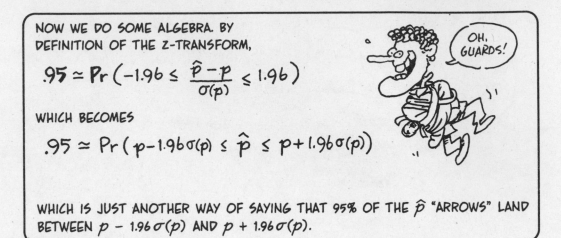

OH, GUARDS!

WHICH IS JUST ANOTHER WAY OF SAYING THAT 95% OF THE $\hat{p}$ "ARROWS" LAND BETWEEN $p - 1.96\,\sigma(p)$ AND $p + 1.96\,\sigma(p)$.

NOW WE'RE IN A POSITION TO VIEW THE TARGET FROM BEHIND! ONE MORE TURN OF THE ALGEBRA CRANK MAKES IT

$$.95 \simeq \Pr\left(\hat{p} - 1.96\,\sigma(p) \leq p \leq \hat{p} + 1.96\,\sigma(p)\right)$$

HERE WE ARE DRAWING CIRCLES AROUND A LOT OF ARROWS (I.E., MAKING INTERVALS AROUND $\hat{p}$) AND SAYING THAT 95% OF THEM COVER p.

BUT THERE IS ONE TINY PROBLEM... **WE DON'T ACTUALLY KNOW THE SIZE OF THE BULL'S-EYE, BECAUSE WE DON'T KNOW p, AND THE WIDTH IS A MULTIPLE OF $\sigma(p)$.**

THE CIRCLES ARE ALL DIFFERENT SIZES NOW, BUT IT'S OKAY, REALLY...

SO WE FUDGE A LITTLE AND USE THE **STANDARD ERROR OF $\hat{P}$:**

$$SE(\hat{P}) = \frac{\sqrt{\hat{p}(1-\hat{p})}}{\sqrt{n}}$$

IN ITS PLACE... IT'S CLOSE ENOUGH... IT'S THE BEST WE CAN DO... AND IT CAN EVEN BE THEORETICALLY JUSTIFIED!

NOW THE FORMULA IS

$$.95 \approx \Pr\left(\hat{p} - 1.96\,SE(\hat{p}) \leq p \leq \hat{p} + 1.96\,SE(\hat{p})\right)$$

AGAIN, THIS EQUATION DESCRIBES THE PROBABILITY THAT THE TRUE, FIXED POPULATION PROPORTION FALLS WITHIN THE *RANDOM INTERVAL*

$$(\hat{p} - 1.96\,SE(\hat{p}),\ \hat{p} + 1.96\,SE(\hat{p})).$$

IF WE SAMPLED REPEATEDLY, THESE INTERVALS WOULD COVER p 95% OF THE TIME.

LET'S STARE AT THIS A MINUTE...

NOW OUR PROBABILITY CALCULATION IS DONE, AND IT'S TIME FOR...

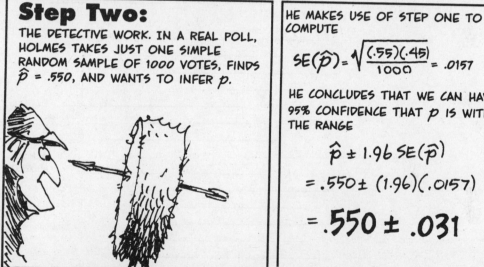

Step Two:

THE DETECTIVE WORK. IN A REAL POLL, HOLMES TAKES JUST ONE SIMPLE RANDOM SAMPLE OF 1000 VOTES, FINDS $\hat{p} = .550$, AND WANTS TO INFER p.

HE MAKES USE OF STEP ONE TO COMPUTE

$$SE(\hat{p}) = \sqrt{\frac{(.55)(.45)}{1000}} = .0157$$

HE CONCLUDES THAT WE CAN HAVE 95% CONFIDENCE THAT p IS WITHIN THE RANGE

$$\hat{p} \pm 1.96\,SE(\hat{p})$$

$$= .550 \pm (1.96)(.0157)$$

$$= .550 \pm .031$$

THIS IS WHAT POLLS MEAN WHEN THEY REFER TO THEIR "MARGIN OF ERROR." IN THIS CASE, HOLMES FOUND THAT

$$.519 \leq p \leq .581,$$

IN OTHER WORDS THAT $p = 55\%$ WITH A 3% MARGIN OF ERROR. (POLLS TYPICALLY USE A 95% CONFIDENCE LEVEL.)

THE MARGIN OF ERROR WAS 3%, WHATEVER *THAT* MEANS...

119

THIS PAGE SHOWS THE RESULTS OF A COMPUTER SIMULATION OF TWENTY
SAMPLES OF SIZE n = 1000. WE ASSUMED THAT THE TRUE VALUE OF p = .5. AT
THE TOP YOU SEE THE SAMPLING DISTRIBUTION OF $\hat{p}$ (NORMAL, WITH MEAN p
AND $\sigma = \sqrt{\frac{p(1-p)}{n}}$). BELOW ARE THE 95% CONFIDENCE INTERVALS FROM EACH
SAMPLE. ON AVERAGE, ONE OUT OF TWENTY (OR 5%) OF THESE INTERVALS WILL
NOT COVER THE POINT p = .5.

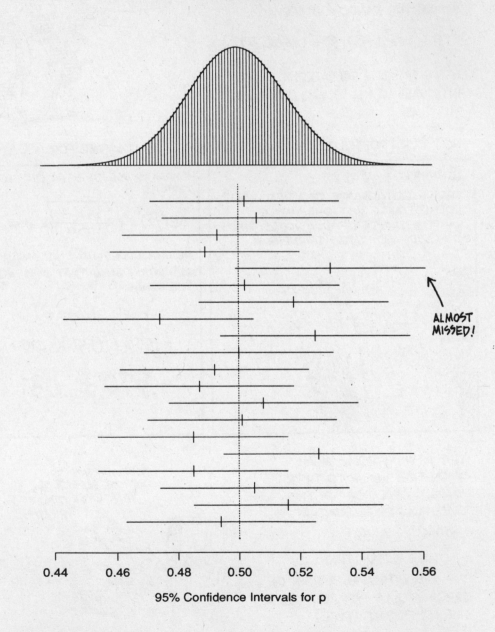

ALMOST
MISSED!

Sample

95% Confidence Intervals for p

0.44   0.46   0.48   0.50   0.52   0.54   0.56

ALTHOUGH 95% CONFIDENCE IS GOOD ENOUGH FOR NEWSPAPER POLLS, IT ISN'T GOOD ENOUGH FOR **SENATOR ASTUTE.** HE WANTS 99%!

ANYTHING LESS, AND MY BIG MONEY PEOPLE WON'T INVEST—I MEAN **CONTRIBUTE** TO MY FIGHT FOR LIBERTY AND JUSTICE!

HOW TO INCREASE CONFIDENCE? USING THE ARCHERY TARGET, WE CAN SEE TWO WAYS: ONE IS TO **INCREASE THE SIZE OF THE CIRCLE** YOU DRAW...

THAT *①#≢ OUGHTA DO IT!

AND ANOTHER WOULD BE TO **IMPROVE THE AIM** OF THE ARCHER IN THE FIRST PLACE, SO HER ARROWS LAND CLOSER TO THE BULL'S-EYE.

THE FIRST METHOD IS EQUIVALENT TO **WIDENING THE CONFIDENCE INTERVAL.** THE GREATER THE MARGIN OF ERROR, THE MORE CERTAIN YOU ARE THE TRUE VALUE OF p LIES IN THE INTERVAL.

I'M 100% CONFIDENT THAT p IS BETWEEN 0 AND 1 !!

MAYBE IT'S TIME TO SEE EXACTLY HOW WE FIND THE ENDS OF THESE CONFIDENCE INTERVALS...

THE RELEVANT NUMBER HERE WE USUALLY CALL α. IT MEASURES THE DIFFERENCE BETWEEN THE DESIRED CONFIDENCE LEVEL AND CERTAINTY. FOR EXAMPLE, WHEN THE CONFIDENCE LEVEL IS 95%, OR 0.95, α IS .05. SO WE SPEAK OF THE $(1-\alpha) \cdot 100\%$ CONFIDENCE LEVEL.

FINDING THE $(1-\alpha) \cdot 100\%$ CONFIDENCE INTERVAL MEANS: LOOK AT A STANDARD NORMAL CURVE, AND FIND THE POINTS $\pm Z$ BETWEEN WHICH THE AREA IS $1-\alpha$.

AREA = .95

-Z   0   Z

THIS POINT, CALLED $z_{\frac{\alpha}{2}}$, IS THE Z-VALUE BEYOND WHICH THE AREA IS $.025 = \frac{\alpha}{2}$.

AREA = .025

0    $z_{\alpha/2}$

THAT'S BECAUSE WE'RE CHOPPING OFF "TAILS" AT BOTH ENDS OF THE CURVE, WHICH HAVE A TOTAL AREA OF $\frac{\alpha}{2} + \frac{\alpha}{2} = \alpha$.

AREA = .025   AREA = .025

$-z_{\alpha/2}$ 0 $z_{\alpha/2}$

WE CAN FIND $z_{\alpha/2}$ STRAIGHT FROM THE STANDARD NORMAL TABLE (PAGE 84). IT'S THE POINT WITH THE PROPERTY

$$Pr(z \geq z_{\alpha/2}) = \frac{\alpha}{2}$$

IN PARTICULAR,

$$Pr(z \geq z_{.025}) = .025$$

z	-2.5	-2.4	-2.3	-2.2	-2.1
F(z)	0.006	0.008	0.011	0.014	0.018
z	-2.0	-1.9	-1.8	-1.7	-1.6
F(z)	0.023	0.029	0.036	0.045	0.055
z	-1.5				
F(z)	0.067	0.0			

$-z_{.025}$ IS IN THIS INTERVAL!

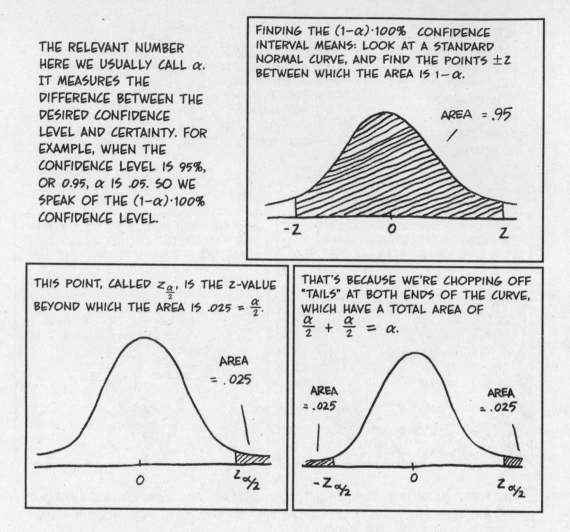

122

HERE'S A LITTLE TABLE OF THE CRITICAL VALUES FOR VARIOUS LEVELS OF CONFIDENCE...

FOR THIS LEVEL OF CONFIDENCE, GO OUT THIS MANY STANDARD DEVIATIONS!

YAWN... JUST THE ANSWER, PLEASE...

$1-\alpha$	.80	.90	.95	.99
α	.20	.10	.05	.01
$\alpha/2$	.10	.05	.025	.005
$z_{\frac{\alpha}{2}}$	1.28	1.64	1.96	2.58

TO MAKE A 99% CONFIDENCE INTERVAL, WE USE THAT TABLE TO WRITE

$$.99 = Pr(\hat{p} - 2.58 SE(\hat{p}) \leqslant p \leqslant \hat{p} + 2.58 SE(\hat{p}))$$

WHICH WE SLOPPILY ABBREVIATE AS

$$p = \hat{p} \pm 2.58 \sqrt{\frac{\hat{p}(1-\hat{p})}{n}}$$

$$= .55 \pm 2.58 \sqrt{\frac{(.55)(.45)}{1000}}$$

$$= .55 \pm .041$$

WITH 99% CONFIDENCE.

GREAT! I'M STILL OVER 50%!

.50   .51   .55   .59

WIDENING THE INTERVAL IS ONE WAY TO INCREASE OUR CONFIDENCE IN THE RESULT. AS WE MENTIONED, ANOTHER WAY WOULD BE TO SHOOT OUR ARROWS *MORE ACCURATELY.* IF WE KNEW THAT THE ARCHER GOT 95% OF HER ARROWS WITHIN *1 CM* OF THE BULL'S-EYE, OUR ESTIMATES COULD BE A LOT SHARPER!

HOW DO WE DO THIS? BY INCREASING THE SAMPLE SIZE! THE WIDTH OF THE CONFIDENCE INTERVAL DEPENDS ON THE SAMPLE SIZE: THE INTERVAL HAS THE FORM $\hat{p} + E$, WHERE E, THE ERROR, IS GIVEN BY

$$E = z_{\frac{\alpha}{2}} \sqrt{\frac{\hat{p}(1-\hat{p})}{n}}$$

SO THE BIGGER WE MAKE n, THE SMALLER THE ERROR. (E.G., QUADRUPLING n HALVES THE INTERVAL WIDTH.)

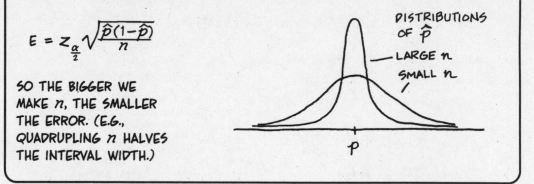

DISTRIBUTIONS OF $\hat{p}$

LARGE n

SMALL n

p

ASTUTE ASKS HOLMES TO GIVE HIM A SMALL ERROR WITH HIGH CONFIDENCE—SAY 99% CONFIDENCE WITH E = ±.01. HOLMES SOLVES FOR n.

$$n = \frac{z_{\frac{\alpha}{2}}^2 \; p^*(1-p^*)}{E^2}$$

(WHERE p^* IS A GUESS AT THE TRUE PROPORTION p—REMEMBER, WE HAVEN'T TAKEN THE SAMPLE YET!)

GET OUT YOUR WALLET, BOSS, I'VE GOT AN ANSWER!

WHAT HAPPENED IS THAT POLITICIANS ARE NOT ELECTED BY POLLS!

OUTRAGEOUS! I'D PASS A LAW AGAINST THIS, IF I WERE STILL IN THE SENATE!

SOME PROBLEMS WITH POLLS, AS OPPOSED TO ELECTIONS:

A RESPONSE BIAS: VOTERS MAY LIE TO THE INTERVIEWER OR CHANGE THEIR MINDS BEFORE ELECTION DAY.

I LOVE BOTH MAJOR PARTIES AND ONLY WISH I COULD VOTE FOR BOTH OF THEM!

YOU CREDULOUS DORK

B ALTHOUGH THE POLL IS AN UNBIASED SAMPLE OF POTENTIAL VOTERS, THE VOTING BOOTH COUNTS ONLY ACTUAL VOTERS.

BUT WASN'T THE ELECTION YESTERDAY?

C NON-RESPONSE BIAS: THE VOTER MAY NOT BE HOME OR REFUSE TO TAKE PART IN THE POLL.

SLAM

THERE IS *NO WAY* FOR A POLLSTER TO GET INSIDE A POTENTIAL VOTER'S HEAD AND KNOW IF SHE'S GOING TO VOTE, IF SHE'S LYING, OR IF SHE'S GOING TO CHANGE HER MIND BEFORE ELECTION DAY. LARGE SAMPLE SIZES CANNOT REDUCE THESE KINDS OF ERRORS.

UM, SORRY...

NEXT TIME, HIRE A PSYCHIC!

126

SINCE THESE ERRORS CAN BE LARGE, IT SELDOM PAYS TO TAKE A VERY LARGE RANDOM SAMPLE.

INSTEAD, WE USE THIS BIAS SQUEEZER!

IN THE LAST FIVE PRESIDENTIAL ELECTIONS, THE GALLUP POLL HAS INTERVIEWED FEWER THAN 4,000 VOTERS FOR EACH ELECTION. YET IN ALL FIVE ELECTIONS, THE GALLUP ORGANIZATION'S ERRORS IN PREDICTING THE PRESIDENTIAL ELECTION OUTCOME HAVE BEEN LESS THAN 2%.

AND WHAT'S THAT?

INDUSTRIAL STRENGTH BIAS SQUEEZER!

THEIR SUCCESS IS DUE TO THEIR USE OF ESTIMATORS THAT ACCOUNT FOR NON-RESPONSE, AND THEY SCREEN OUT ELIGIBLE VOTERS WHO ARE NOT LIKELY TO VOTE.

WHAT ABOUT THESE?

OUTSIDE OF TEXAS AND CHICAGO, I THINK WE'RE SAFE...

TO SUMMARIZE, ESTIMATED PROPORTION = TRUE PROPORTION + BIAS + RANDOM SAMPLING ERROR. EVEN POLLSTERS HAVE LIMITED FUNDS. THEY WISELY CHOOSE TO SPEND THEIR MONEY *REDUCING BIAS*, RATHER THAN INCREASING THE SAMPLES BEYOND 4,000 VOTERS.

Confidence Intervals for μ

UP TO NOW, WE'VE BEEN LOOKING AT CONFIDENCE INTERVALS FOR A PROPORTION p OF A POPULATION. EXACTLY THE SAME REASONING WORKS FOR THE POPULATION MEAN μ.

IN THE LAST CHAPTER (P. 105), WE SAW THAT THE DISTRIBUTION OF SAMPLE MEANS $\overline{X}$ IS APPROXIMATELY **NORMAL**, CENTERED ON THE ACTUAL POPULATION MEAN μ, WITH STANDARD DEVIATION $\sigma/\sqrt{n}$, WHERE σ IS THE POPULATION STANDARD DEVIATION. SO, FOR LARGE n,

$$.95 = Pr(-1.96 \leq Z \leq 1.96)$$

$$\approx Pr\left(-1.96 \leq \frac{\overline{X}-\mu}{\sigma/\sqrt{n}} \leq 1.96\right)$$

AGAIN, NOT KNOWING σ, WE REPLACE σ WITH S, THE SAMPLE STANDARD DEVIATION:

$$.95 \approx Pr\left(-1.96 \leq \frac{\overline{X}-\mu}{s/\sqrt{n}} \leq 1.96\right)$$

THE TERM $s/\sqrt{n}$ IS CALLED THE **SAMPLE STANDARD ERROR**, AND WRITTEN SE($\overline{X}$). WE CONCLUDE THAT

$$.95 \approx Pr(\overline{X}-1.96\,SE(\overline{X}) \leq \mu \leq \overline{X}+1.96\,SE(\overline{X}))$$

WHERE

$$SE(\overline{X}) = \frac{s}{\sqrt{n}}$$

JUST AS BEFORE, WE HAVE
FOUND THAT THE RANDOM
INTERVAL

$$\bar{X} \pm 1.96\,SE(\bar{X})$$

COVERS THE TRUE MEAN, μ, WITH
PROBABILITY .95... SO NOW WE CAN
CALL IN SHERLOCK HOLMES TO
MAKE A STATISTICAL INFERENCE
BASED ON A SINGLE SAMPLE OF
SIZE n WITH MEAN $\bar{x}$.

PIECE OF
CAKE!

HE (AND WE) ARE 95% CONFIDENT THAT THE MEAN μ IS WITHIN THE INTERVAL
$\bar{x} \pm 1.96\,SE(\bar{X})$.

BY GAD, I'M
GETTING MORE
CONFIDENT WITH
EVERY PASSING
MOMENT!

AS BEFORE, FOR AN ARBITRARY
LEVEL OF CONFIDENCE $1-\alpha$,
WE REPLACE 1.96 BY $z_{\frac{\alpha}{2}}$.

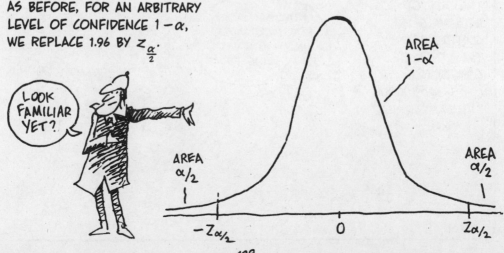

LOOK
FAMILIAR
YET?

AREA
$1-\alpha$

AREA
$\alpha/2$

AREA
$\alpha/2$

$-z_{\alpha/2}$

0

$z_{\alpha/2}$

LET'S REVISIT THE STUDENT WEIGHT DATA FROM CHAPTER 2, ASSUMING THAT THE $n = 92$ STUDENTS WERE A SIMPLE RANDOM SAMPLE OF ALL PENN STATE STUDENTS.

QUICK! SOMEONE DIVIDE BY n!

THE SAMPLE MEAN $\bar{x}$ WAS 145.2 LBS. AND SAMPLE STANDARD DEVIATION S WAS 23.7. SO THE STANDARD ERROR IS

$$SE(\bar{x}) = \frac{23.7}{\sqrt{92}} = \mathbf{2.47}$$

AND WE NOW HAVE 95% CONFIDENCE THAT THE MEAN WEIGHT OF ALL PENN STATE STUDENTS FALLS IN THE INTERVAL

$$\bar{x} \pm 1.96 SE(\bar{x})$$
$$= 145.2 \pm (1.96)(2.47)$$
$$= \mathbf{145.2 \pm 4.8 \ POUNDS}$$

TO SUMMARIZE: FOR A SIMPLE RANDOM SAMPLE (SRS) OF LARGE SIZE, THE $(1-\alpha) \cdot 100\%$ CONFIDENCE INTERVAL IS:

POPULATION MEAN, μ

$$\mu = \bar{x} \pm z_{\frac{\alpha}{2}} SE(\bar{x})$$

WHERE $SE(\bar{x}) = {s}/{\sqrt{n}}$

POPULATION PROPORTION, p

$$p = \hat{p} \pm z_{\frac{\alpha}{2}} SE(\hat{p})$$

WHERE $SE(\hat{p}) = \sqrt{\dfrac{\hat{p}(1-\hat{p})}{n}}$

THE SIZE OF BOTH INTERVALS IS CONTROLLED BY THE LEVEL OF CONFIDENCE $(1-\alpha) \cdot 100\%$ AND THE SAMPLE SIZE, n.

NOW, SENATOR, HOW WOULD YOU LIKE A *JOB* WITH MY POLLING FIRM?

Student's t (again!)

AS WE SAW IN CHAPTER 6, THE STATISTIC

$$\frac{\bar{X} - \mu}{SE(\bar{X})}$$

HAS AN APPROXIMATELY NORMAL DISTRIBUTION ONLY WHEN IT IS COMPUTED USING A **LARGE SAMPLE**. FOR SMALL SAMPLES ($n=5, 10, 25...$), THIS IS NO LONGER THE CASE, AND WE HAVE TO USE THE STUDENT'S t.

LET'S LOOK AT t A LITTLE MORE CLOSELY. WE MENTIONED THAT THE t DISTRIBUTION IS MORE SPREAD OUT THAN THE NORMAL, AND THAT THE AMOUNT OF SPREAD DEPENDS ON THE SAMPLE SIZE.

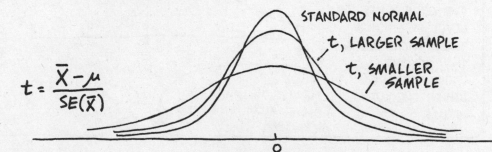

$$t = \frac{\bar{X} - \mu}{SE(\bar{X})}$$

STANDARD NORMAL

t, LARGER SAMPLE

t, SMALLER SAMPLE

0

WHAT ITS DISCOVERER GOSSET DID WAS TO **QUANTIFY** THIS RELATIONSHIP. IF n IS THE SAMPLE SIZE, HE SAID, THEN CALL $n-1$ THE NUMBER OF

degrees of freedom

OF THE SAMPLE.

THE GENERAL IDEA: GIVEN n PIECES OF DATA $x_1, x_2, ... x_n$, YOU USE UP ONE "DEGREE OF FREEDOM" WHEN YOU COMPUTE $\bar{x}$, LEAVING $n-1$ INDEPENDENT PIECES OF INFORMATION.

$x_1 \, x_2 \, x_3 \, ... x_n$

$\sum \frac{x_i}{n}$

GOSSET COMPUTED TABLES OF THE t DISTRIBUTION FOR DIFFERENT SAMPLE SIZES—I.E., DEGREES OF FREEDOM. WE REPEAT, THE *MORE DEGREES OF FREEDOM*, THE CLOSER t BECOMES TO THE *STANDARD NORMAL*.

A NICE, SLOPPY DISTRIBUTION!

KNOWING THE SAMPLE SIZE n, WE CHOOSE THE t DISTRIBUTION WITH $n-1$ DEGREES OF FREEDOM.

AS WITH THE z DISTRIBUTION (I.E., THE STANDARD NORMAL), WE GET A 95% CONFIDENCE LEVEL BY FINDING THE CRITICAL VALUE $t_{.025}$ BEYOND WHICH THE AREA UNDER THE CURVE IS .025.

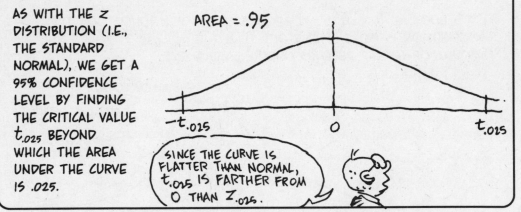

AREA = .95

$-t_{.025}$ 0 $t_{.025}$

SINCE THE CURVE IS FLATTER THAN NORMAL, $t_{.025}$ IS FARTHER FROM 0 THAN $z_{.025}$.

FOR A $(1-\alpha)\cdot100\%$ CONFIDENCE INTERVAL, WE FIND THE CRITICAL VALUE $t_{\frac{\alpha}{2}}$ SUCH THAT $Pr(t \geq t_{\frac{\alpha}{2}}) = \frac{\alpha}{2}$. HERE IS A SHORT TABLE OF CRITICAL VALUES FOR THE t DISTRIBUTION:

	$1-\alpha$	.80	.90	.95	.99
	α	.20	.10	.05	.01
	$\alpha/2$	.10	.05	.025	.005
DEGREES OF FREEDOM	1	3.09	6.31	12.71	63.66
	10	1.37	1.81	2.23	4.14
	30	1.31	1.70	2.04	2.75
	100	1.29	1.66	1.98	2.63
	∞	1.28	1.65	1.96	2.58

EACH COLUMN REPRESENTS A FIXED LEVEL OF CONFIDENCE, WITH INCREASING NUMBERS OF DEGREES OF FREEDOM. THE HIGHER THE DEGREES OF FREEDOM, THE CLOSER THE CRITICAL VALUE GETS TO $z_{\alpha/2}$, THE CRITICAL VALUE OF THE NORMAL DISTRIBUTION.

WE DERIVE THE WIDTH OF OUR CONFIDENCE INTERVAL DIRECTLY FROM THE DEFINITION OF t:

$$t = \frac{\overline{X} - \mu}{SE(\overline{X})}$$

NOTE: IT'S *EXACTLY* LIKE THE CASE OF A LARGE SAMPLE, BUT WITH t INSTEAD OF z!

THEN, FOR CONFIDENCE LEVEL $(1-\alpha)\cdot100\%$,

$$(1-\alpha) = Pr\left(\overline{x} - t_{\frac{\alpha}{2}} SE(\overline{X}) \leq \mu \leq \overline{x} + t_{\frac{\alpha}{2}} SE(\overline{X})\right)$$

FROM WHICH WE INFER: GIVEN A SINGLE SAMPLE OF SIZE n AND MEAN $\overline{x}$, WE CAN BE $(1-\alpha)\cdot100\%$ CONFIDENT THAT THE POPULATION MEAN μ FALLS IN THE RANGE

$$\mu = \overline{x} \pm t_{\frac{\alpha}{2}} SE(\overline{x})$$

WHERE $SE(\overline{x}) = \frac{s}{\sqrt{n}}$ AND $t_{\frac{\alpha}{2}}$ IS THE CRITICAL VALUE OF THE t DISTRIBUTION WITH $n-1$ DEGREES OF FREEDOM.

YOU· WILL· MEMORIZE· THIS...

STILL AWAKE?

NOTE: STRICTLY SPEAKING, THE DERIVATION OF THE t DISTRIBUTION DEPENDED ON THE ASSUMPTION THAT THE SAMPLE WAS FROM A NORMAL POPULATION. IN PRACTICE, CONFIDENCE INTERVALS BASED ON THE t WORK REASONABLY WELL, EVEN WHEN THE POPULATION DISTRIBUTION IS ONLY APPROXIMATELY MOUND-SHAPED.

example: SUPPOSE *CHAMELEON MOTORS* HAS TO CRASH TEST ITS CARS TO DETERMINE THE AVERAGE REPAIR COST OF A 10 M.P.H. HEAD-ON COLLISION. THIS IS EXPENSIVE! THEY DECIDE TO TRY IT ON JUST *FIVE* CHAMELEONS.

THEY FIND THE DAMAGE DATA TO BE $150, $400, $720, $500, AND $930.

THE SAMPLE MEAN:

$$\bar{x} = \$540$$

THE STANDARD DEVIATION:

$$S = \$299$$

YOU CAN CHECK S WITH A HAND CALCULATOR. IT'S

$$\sqrt{\frac{1}{4}\left((150-540)^2 + (400-540)^2 + (720-540)^2 + (500-540)^2 + (930-540)^2\right)}$$

SO WHERE CAN WE PLACE THE MEAN WITH 95% CONFIDENCE? WE FIND OUR CRITICAL VALUE $t_{.025}$ WITH 4 DEGREES OF FREEDOM:

	$1-\alpha$	.80	.90	.95	.99
	α	.20	.10	.05	.01
	$\alpha/2$	.10	.05	.025	.005
DEGREES OF FREEDOM	1	3.09	6.31	12.71	63.66
	2	1.89	2.92	4.30	9.92
	3	1.64	2.35	3.18	5.84
	4	1.53	2.13	2.78	4.60
	5	1.48	2.01	2.57	4.03

AND PLUG IT IN:

$$\mu = \bar{x} \pm 2.78 \, \frac{s}{\sqrt{n}}$$

$$= 540 \pm 2.78 \left(\frac{299}{\sqrt{5}} \right)$$

$$= \mathbf{540 \pm 372}$$

SO THE BEST WE CAN SAY WITH 95% CONFIDENCE IS THAT THE AVERAGE DAMAGE WILL LIE BETWEEN $168 AND $912.

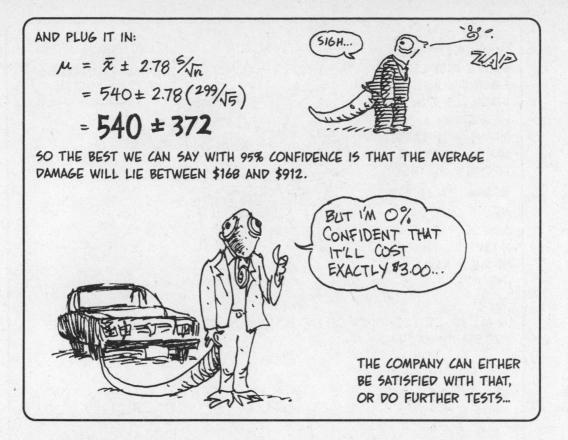

SIGH... ZAP

BUT I'M 0% CONFIDENT THAT IT'LL COST EXACTLY $3.00...

THE COMPANY CAN EITHER BE SATISFIED WITH THAT, OR DO FURTHER TESTS...

TO COMPUTE THIS CONFIDENCE INTERVAL USING STUDENT'S t, WE HAVE MADE AN *UNSTATED ASSUMPTION*: WE ASSUMED THAT CRASH REPAIR COSTS ARE APPROXIMATELY *NORMALLY DISTRIBUTED*, I.E., IF WE CRASHED 1000 CHAMELEONS, THE HISTOGRAM OF REPAIR COSTS WOULD BE SYMMETRICAL AND MOUND-SHAPED. WE *CAN NOT KNOW THIS* FROM 5 DATA POINTS ALONE... BUT MAYBE YEARS OF EXPERIENCE WITH EARLIER MODELS PROVIDE NORMALLY DISTRIBUTED COST HISTOGRAMS FOR FRONT END REPAIRS: INFORMATION WHICH WOULD TEND TO SUPPORT OUR USE OF STUDENT'S t.

AND REAR-ENDERS?

THE TAIL GROWS BACK BY ITSELF. IT'S A FEATURE ON CHAMELEONS..

TO SUM UP (!), WE NOW HAVE THREE SIMPLE RECIPES FOR FINDING CONFIDENCE INTERVALS. FOR PROPORTIONS, OR MEANS WITH LARGE SAMPLE SIZES, WE LOOK UP $z_{\frac{\alpha}{2}}$ IN A NORMAL TABLE. FOR MEANS OF SMALL SAMPLE SIZES (SAY $n \leq 30$), WE FIND $t_{\frac{\alpha}{2}}$ IN THE t TABLE.

SITTING AT A t·TABLE, READING UP ON z·TABLES...

IN ALL CASES, THE WIDTH OF THE INTERVAL IS THAT CRITICAL VALUE TIMES THE STANDARD ERROR:

$$z_{\frac{\alpha}{2}}SE(\hat{p}) \qquad z_{\frac{\alpha}{2}}SE(\overline{X}) \qquad t_{\frac{\alpha}{2}}SE(\overline{X})$$

AND EACH OF THOSE STANDARD ERRORS IS PROPORTIONAL TO THAT MAGIC NUMBER:

$$\frac{1}{\sqrt{n}}$$

◆Chapter 8◆
HYPOTHESIS TESTING

NOW WE ENTER A NEW AREA... GOVERNMENT, BUSINESS, AND THE HARD AND SOFT SCIENCES ALL USE AND OFTEN ABUSE THESE TESTS OF SIGNIFICANCE. IT'S ALL ABOUT ANSWERING THE QUESTION, *"COULD THESE OBSERVATIONS REALLY HAVE OCCURRED BY CHANCE?"*

WE BEGIN WITH AN EXAMPLE FROM THE LAW: A COMPOSITE OF SEVERAL CASES ARGUED IN THE SOUTH BETWEEN 1960 AND 1980, IN WHICH EXPERT WITNESSES PRESENTED THE CASE FOR *RACIAL BIAS IN JURY SELECTION.*

PURE COINCIDENCE!

PANELS OF JURORS ARE THEORETICALLY DRAWN AT RANDOM FROM A LIST OF ELIGIBLE CITIZENS. HOWEVER, IN SOUTHERN STATES IN THE '50S AND '60S, FEW AFRICAN AMERICANS WERE FOUND ON JURY PANELS, SO SOME DEFENDANTS CHALLENGED THE VERDICTS. ON APPEAL, AN EXPERT STATISTICAL WITNESS GAVE THIS EVIDENCE:

1) 50% OF ELIGIBLE CITIZENS WERE AFRICAN AMERICAN.

2) ON AN 80-PERSON PANEL OF POTENTIAL JURORS, ONLY *FOUR* WERE AFRICAN AMERICANS.

COULD THIS BE THE RESULT OF *PURE CHANCE?*

FOR THE SAKE OF ARGUMENT, SUPPOSE THAT THE SELECTION OF POTENTIAL JURORS WAS *RANDOM*. THEN THE NUMBER OF AFRICAN AMERICANS ON THE 80-PERSON PANEL WOULD BE THE *BINOMIAL* RANDOM VARIABLE X WITH $n = 80$ TRIALS AND $p = .5$.

80 BERNOULLI TRIALS, EACH WITH $p = \frac{1}{2}$!

THUS, THE CHANCES OF GETTING A JURY WITH ONLY 4 AFRICAN AMERICANS IS $\Pr(X \leq 4)$, WHICH WORKS OUT TO ABOUT .000000000000000000014 (!).

IS THAT A SMALL NUMBER OR A BIG NUMBER?

THIS IS A DEDUCTIVE PROBABILITY ARGUMENT.

SINCE THE PROBABILITY IS SO SMALL, THE PARTICULAR PANEL WITH ONLY FOUR BLACK MEMBERS IS *STRONG EVIDENCE* AGAINST THE *HYPOTHESIS* OF RANDOM SELECTION.

RANDOM? I ASK YOU!

TO DRIVE THE POINT HOME, THE STATISTICIAN NOTES THAT THIS PROBABILITY IS LESS THAN THE CHANCES OF GETTING *THREE CONSECUTIVE ROYAL FLUSHES* IN POKER.

GASP

SO THE JUDGE *REJECTS THE HYPOTHESIS* OF RANDOM SELECTION.

IF I WAS IN THAT POKER GAME, I'D A STARTED SHOOTIN' AFTER THE SECOND ROYAL FLUSH...

(AND ORDERS HIS OWN REMARKS STRICKEN FROM THE RECORD!)

LET'S FOLLOW THE PROCESS AGAIN TO
SORT OUT THE *FOUR FORMAL STEPS* OF
STATISTICAL HYPOTHESIS TESTING.

Step 1. FORMULATE ALL HYPOTHESES.

H_o, THE *NULL HYPOTHESIS*, IS
USUALLY THAT THE
OBSERVATIONS ARE THE RESULT
PURELY OF *CHANCE*.

H_a, THE *ALTERNATE HYPOTHESIS*,
IS THAT THERE IS A REAL
EFFECT, THAT THE
OBSERVATIONS ARE THE
RESULT OF THIS REAL EFFECT,
PLUS CHANCE VARIATION.

IN THE COURT CASE, H_o SAYS THE
JURY WAS *RANDOMLY CHOSEN*
FROM THE WHOLE POPULATION.
AFRICAN AMERICANS HAVE
PROBABILITY $p = .50$ OF BEING
CHOSEN..

H_a SAYS THAT AFRICAN AMERICANS
ARE LESS LIKELY THAN THEIR
PROPORTION IN THE POPULATION
TO BE SELECTED FOR A JURY
PANEL: $p < .50$.

Step 2. THE *TEST STATISTIC*.
IDENTIFY A STATISTIC THAT WILL ASSESS
THE EVIDENCE AGAINST THE NULL
HYPOTHESIS.

IN THE COURT CASE, THE TEST
STATISTIC IS THE BINOMIAL RANDOM
VARIABLE X WITH $p = .50$ AND
$n = 80$.

Step 3. P-VALUE:

A PROBABILITY STATEMENT WHICH ANSWERS THE QUESTION: IF THE NULL HYPOTHESIS WERE TRUE, THEN WHAT IS THE PROBABILITY OF OBSERVING A TEST STATISTIC AT LEAST AS EXTREME AS THE ONE WE OBSERVED?

THE SMALLER THE P-VALUE, THE STRONGER THE EVIDENCE AGAINST H_0.

IN THE EXAMPLE, THE P-VALUE WAS

$$Pr(x \leq 4 \mid p = .50 \text{ AND } n = 80)$$

$$= 1.4 \times 10^{-18}$$

WE COMPUTED THIS P-VALUE THE MODERN WAY, USING A STATISTICAL SOFTWARE PACKAGE.

IN THE '50s, WE USED HORSE-DRAWN COMPUTERS!

Step 4. COMPARE THE P-VALUE TO A FIXED *SIGNIFICANCE LEVEL*, α.

α ACTS AS A CUT-OFF POINT BELOW WHICH WE AGREE THAT AN EFFECT IS STATISTICALLY SIGNIFICANT. THAT IS, IF

$$\text{P-VALUE} \leq \alpha$$

THEN WE *RULE OUT THE NULL HYPOTHESIS* H_0 AND AGREE THAT SOMETHING ELSE IS GOING ON.

FISHY FISHY FISHY!

IN THE JURY CASE, THE STATISTICIAN TOOK α TO BE 3.6×10^{-18}, THE CHANCES OF BEING DEALT THREE ROYAL FLUSHES IN A ROW.

A P-VALUE EVEN A JUDGE CAN UNDERSTAND!

IN SCIENTIFIC WORK, A FIXED α-LEVEL OF .05 OR .01 IS OFTEN USED. THESE FIXED LEVELS ARE A HOLDOVER ARTIFACT FROM THE PRE-COMPUTER ERA, WHEN WE HAD TO REFER TO TABLES, WHICH WERE PRINTED ONLY FOR SELECTED CRITICAL VALUES. STILL, MANY SCIENTIFIC JOURNALS CONTINUE TO PUBLISH RESULTS ONLY WHEN THE P-VALUE $\leq$ .05.

IN LEGAL PROCEEDINGS, THE STANDARD IS MORE FLEXIBLE...

LARGE SAMPLE
SIGNIFICANCE TEST FOR PROPORTIONS

THE JURY EXAMPLE WAS A SPECIAL CASE OF A GENERAL PROBLEM. THE NULL HYPOTHESIS HAD THE FORM $p = p_0$, WHERE p_0 WAS SOME PROBABILITY (IN THIS CASE, .5), NOW LET'S LOOK AT SUCH PROBLEMS GENERALLY: LET'S *TEST THE HYPOTHESIS* $p = p_0$.

AS USUAL, WE IMAGINE WE HAVE A BIG POPULATION... WE OBSERVE A LARGE SAMPLE... AND WE FIND THAT SOME CHARACTERISTIC OCCURS WITH PROBABILITY $\hat{p}$.

BASED ON THIS OBSERVATION, WE WANT TO KNOW IF THE TRUE POPULATION PROBABILITY IS (FOR INSTANCE) LARGER THAN SOME OTHER VALUE p_0. FOR EXAMPLE, SENATOR ASTUTE, HAVING FOUND A $\hat{p}$ OF .55, WOULD LIKE TO KNOW THAT $p > .5$, A WINNING MAJORITY.

Step 1.

THE *NULL* HYPOTHESIS IS

$$H_0 : p = p_0$$

THE *ALTERNATE* HYPOTHESIS DEPENDS ON THE DIRECTION OF THE EFFECT WE ARE LOOKING FOR. IN SENATOR ASTUTE'S CASE,

$$H_a : p > p_0$$

BUT IN OTHER CASES, THE ALTERNATE HYPOTHESIS MIGHT WELL BE

$$H_a : p < p_0$$
OR
$$H_a : p \neq p_0$$

FOR EXAMPLE, IN THE JURY SELECTION EXAMPLE, THE ALTERNATIVE HYPOTHESIS WAS

$$H_a : p < 0.5$$

AND AT OTHER TIMES, WE ARE INTERESTED IN KNOWING THAT p IS DIFFERENT FROM SOME VALUE p_0. FOR INSTANCE, IN TESTING FOR A FAIR COIN, WE HAVE AN ALTERNATE HYPOTHESIS OF

$$H_a : p \neq 0.5$$

BUT HAVE NO *A PRIORI* OPINION ABOUT WHETHER HEADS OR TAILS WILL COME UP MORE OFTEN.

Step 2. THE TEST STATISTIC IS

$$z_{OBS} = \frac{\hat{p} - p_0}{\sqrt{p_0(1-p_0)}/\sqrt{n}}$$

WHICH MEASURES HOW FAR p DEVIATES FROM p_0. UNDER THE NULL HYPOTHESIS, z_{OBS} HAS THE STANDARD NORMAL DISTRIBUTION.

Step 3. THE P-VALUE DEPENDS ON WHICH ALTERNATE HYPOTHESIS IS RELEVANT:

a) "RIGHT-HANDED" $H_a : p > p_0$ USES P-VALUE $Pr(z > z_{OBS})$

b) "LEFT-HANDED" $H_a : p < p_0$ USES P-VALUE $Pr(z < z_{OBS})$

c) "TWO-SIDED" $H_a : p \neq p_0$ USES P-VALUE $Pr(|z| > |z_{OBS}|)$

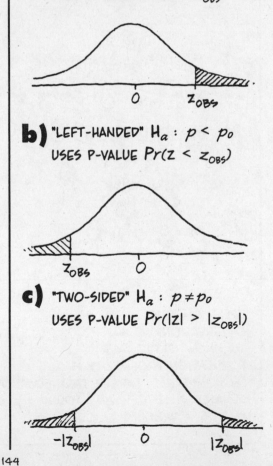

IN THE CASE OF SENATOR ASTUTE:

1) THE HYPOTHESES ARE

$H_0 : p = .5$

$H_a : p > .5$

2) HIS TEST STATISTIC IS

$$z_{OBS} = \frac{.55 - .50}{\sqrt{(.5)(.5)}/\sqrt{1000}} = 3.16$$

3) HIS P-VALUE IS

$Pr(z > z_{OBS}) = Pr(z \geq 3.16) = .0008$

(FROM THE NORMAL TABLE).

4) ASTUTE, BEING FAIRLY CONSERVATIVE, TAKES A SIGNIFICANCE LEVEL α OF .01 AND OBSERVES THAT

$Pr(z > z_{OBS}) = .0008 < \alpha$

THE SENATOR THUS REJECTS THE NULL HYPOTHESIS, AND HE (AND HIS BACKERS) NOW FEEL CERTAIN HE'S IN THE LEAD.

LARGE SAMPLE
TEST FOR THE
POPULATION MEAN

HERE IS HOW A SIGNIFICANCE TEST MIGHT BE USED IN *INSPECTION SAMPLING,* AN IMPORTANT INDUSTRIAL APPLICATION.

FEELS **LIGHT,** MAN...

HEAVY, MAN...

NEW AGE GRANOLA INC. CLAIMS THAT THE AVERAGE WEIGHT OF ITS CEREAL BOXES IS AT LEAST *16 OZ.* THE *GENUINE GROCERY CORPORATION* WILL SEND BACK A SHIPMENT IF THE AVERAGE WEIGHT IS ANY LESS.

BUT OF COURSE GENUINE GROCERY HAS *NO INTENTION* OF WEIGHING EVERY BOX IN A SHIPMENT. THEY'RE GOING TO USE STATISTICS!

STATISTICS IS THE EASY WAY, REMEMBER?

FIRST, THEY CHOOSE THEIR
HYPOTHESES.

$$H_0: \mu = 16 \ OZ.$$

$$H_a: \mu < 16 \ OZ.$$

REJECTING THE NULL
HYPOTHESIS MEANS
REFUSING THE GRANOLA.

NEXT, THEY CHOOSE A TEST STATISTIC. BY NOW, IT SHOULD BE PRETTY MUCH A
KNEE-JERK REACTION TO KNOW THAT THE SAMPLE SPREAD FROM THE MEAN IS

$$\frac{\bar{X} - \mu_0}{SE(\bar{X})} = \frac{\bar{X} - \mu_0}{s/\sqrt{n}}$$

WHERE s IS THE SAMPLE
STANDARD DEVIATION. UNDER
THE NULL HYPOTHESIS, THIS
APPROXIMATES THE
STANDARD NORMAL WHEN
THE SAMPLE IS LARGE, BY
THE CENTRAL LIMIT
THEOREM.

SKIPPING OVER STEP 3 FOR A MOMENT, THEY SET A SIGNIFICANCE LEVEL. BEING A
BUNCH OF DROPPED-OUT SCIENCE MAJORS, THE GROCERS THINK $\alpha = .05$ SOUNDS
ABOUT RIGHT.

JUST THEN, A BOXCAR
LOADED WITH 10,000
BOXES OF GRANOLA
ARRIVES AT THE DOOR.

THEY PULL OUT A
SIMPLE RANDOM
SAMPLE OF 49 BOXES,
WEIGH EACH ONE, AND
DETERMINE THE
SAMPLE'S SUMMARY
STATISTICS:

$$\bar{x} = 15.90 \text{ oz.}$$
$$S = .35 \text{ oz.}$$

A LITTLE LIGHT—BUT
SIGNIFICANTLY SO?

THEY PLUG THE VALUES INTO THE TEST STATISTIC TO FIND

$$Z_{OBS} = \frac{15.9 - 16}{.35/\sqrt{49}} = -2$$

NOW THEY COMPUTE THE P-VALUE:

$$Pr(z < -2 \mid H_0) = .0227$$

THIS BEING LESS THAN THE .05
SIGNIFICANCE LEVEL, GENUINE GROCERY
REJECTS THE NULL HYPOTHESIS, AND
THE SHIPMENT.

SEND IT
BACK, YOU
BURN
ARTIST!!

☆*❂

NE
CR

"FLAKY FOOD FRO

WHAT
HAPPENED?

PRESIDENT

GOT THE MUNCHIES,
MAN... I DIDN'T
THINK ANYONE WOULD
NOTICE IF I ATE A
LITTLE FROM EVERY
BOX...

SMALL SAMPLE
TEST FOR THE POPULATION
MEAN

ARE YOU INSURED FOR THIS?

WE RETURN TO CHAMELEON MOTORS, AND ITS 10 M.P.H. CRASH TEST. THE **RIGHTEOUS INSURANCE COMPANY** WILL INSURE AN AUTO ONLY IF THE MEAN REPAIR COST AFTER A 10 M.P.H. COLLISION IS LESS THAN $1000. THE COMPANY USES A STANDARD $\alpha = .05$ AS ITS SIGNIFICANCE LEVEL. SO...

H_0: $\mu \geq \$1000$ MEAN COST IS TOO HIGH

H_a: $\mu < \$1000$ MEAN COST IS O.K.

THE TEST STATISTIC IS THE t DISTRIBUTION

$$t = \frac{\overline{X} - \mu_0}{SE(\overline{X})}$$

WHERE μ_0 IS THE HYPOTHETICAL MEAN OF $1000

t, 4 DEGREES OF FREEDOM

$-t_{.05}$ O

AND WE WANT OUR OBSERVED t VALUE TO LIE TO THE LEFT OF $-t_{.05}$ (BECAUSE LOW $\overline{x}$ IS DESIRABLE, $\overline{x} - \mu_0$ SHOULD BE NEGATIVE, TO SUPPORT H_a).

	α		
	.05	.025	.005
DEGREES OF FREEDOM 1	6.31	12.71	63.66
2	2.92	4.30	9.92
3	2.35	3.18	5.84
4	2.13	2.78	4.60
5	2.01	2.57	4.03

FROM THE TABLE OF CRITICAL t VALUES, WE SEE THAT $t_{.05} = 2.13$, SO WE DECIDE TO REJECT H_0 IF

$$t_{OBS} \leq -t_{.05} = -2.13$$

FROM CHAPTER 8, WE HAVE $\bar{x} = \$540$ AND $S = \$299$ FOR A SMALL, FIVE-CAR SAMPLE, SO WE FIND

$$t_{OBS} = \frac{540-1000}{299/\sqrt{5}}$$

$$= -3.44 < -t_{.05}$$

THE CAR PASSES THE TEST... H_0 IS REJECTED... AND THE INSURANCE POLICY IS ISSUED.

THIS IS AN EXAMPLE OF **ACCEPTANCE SAMPLING**. THE NULL HYPOTHESIS IS THAT REPAIR COSTS ARE UNACCEPTABLE, AND THE MOTOR COMPANY IS ASSUMED GUILTY UNTIL IT PRESENTS SUFFICIENT EVIDENCE OF ITS INNOCENCE—I.E., THAT ITS PRODUCT IS WITHIN SPECIFICATIONS.

DECISION THEORY

WE CAN THINK OF HYPOTHESIS TESTING AND
SIGNIFICANCE TESTS IN TERMS OF A *HOUSEHOLD
SMOKE-DETECTOR*. IF YOU HAVE ONE OF THESE
WHERE YOU LIVE, YOU'VE PROBABLY NOTICED HOW IT
TENDS TO GO OFF EVERY TIME YOU MAKE THE TOAST
TOO DARK!

THIS IS WHAT IS CALLED A *TYPE I ERROR*: AN ALARM WITHOUT A FIRE.
CONVERSELY, A *TYPE II ERROR* IS A FIRE WITHOUT AN ALARM. EVERY COOK
KNOWS HOW TO AVOID A TYPE I ERROR: JUST *REMOVE THE BATTERIES*.
UNFORTUNATELY, THIS INCREASES THE INCIDENCE OF TYPE II ERRORS!

SIMILARLY, REDUCING THE CHANCES OF TYPE II ERROR, FOR EXAMPLE BY MAKING
THE ALARM HYPERSENSITIVE, CAN INCREASE THE NUMBER OF FALSE ALARMS.

WE CAN SUMMARIZE THIS IN A TWO-BY-TWO *DECISION TABLE.*

	NO FIRE	FIRE
NO ALARM	NO ERROR	TYPE II
ALARM	TYPE I	NO ERROR

NOW THINK OF THE NULL HYPOTHESIS AS THE CONDITION OF *NO FIRE*, WHILE THE ALTERNATE HYPOTHESIS IS THAT A FIRE IS BURNING. THE ALARM CORRESPONDS TO REJECTION OF THE NULL HYPOTHESIS:

TRUE STATE

	H_0	H_a
ACCEPT H_0	NO ERROR	TYPE II
REJECT H_0	TYPE I	NO ERROR

ALL THE SIGNIFICANCE TESTS WE DID EARLIER IN THIS CHAPTER EMPHASIZED THE PROBABILITY OF COMMITTING A TYPE I ERROR—I.E., THE PROBABILITY OF OUR OBSERVATIONS OCCURRING IF H_0 WAS TRUE. WE DEMANDED THAT

$$Pr(\text{REJECTING } H_0 \mid H_0) = Pr(\text{TYPE I ERROR} \mid H_0) = \alpha$$

$1-\alpha$ MEASURES OUR CONFIDENCE THAT ANY ALARM BELLS WE HEAR ARE GENUINE. HIGH CONFIDENCE MEANS RARELY SETTING OFF FALSE ALARMS.

BUT SOMETIMES WHAT WE REALLY WANT TO KNOW IS THE CHANCE OF MAKING A *TYPE II* ERROR! IN OTHER WORDS, HOW SENSITIVE IS OUR "ALARM SYSTEM" WHEN THE ALTERNATE HYPOTHESIS IS *TRUE?*

AN ENVIRONMENTAL EXAMPLE:

IN THE PAST, FACTORIES DISCHARGING CHEMICALS INTO WATERWAYS WERE REQUIRED TO SHOW THAT THE DISCHARGE HAD NO EFFECT ON THE DOWN-STREAM WILDLIFE. THAT'S H_0. THE POLLUTER COULD CONTINUE AS LONG AS THE NULL HYPOTHESIS WAS NOT REJECTED AT THE .05 SIGNIFICANCE LEVEL.

BURBLE BORBLE

SO A POLLUTER, SUSPECTING THAT HE WAS IN VIOLATION OF EPA STANDARDS, WOULD DEVISE AN INEFFECTIVE POLLUTION MONITORING PROGRAM.

WE'LL INTERVIEW A FEW DUCKS!!

BUP BUP

THE POLLUTER IS DELIGHTED, SINCE, LIKE OUR SMOKE ALARM WITHOUT A BATTERY, HIS TEST HAS LITTLE OR NO CHANCE OF SETTING OFF AN ALARM.

LET'S FORMALIZE THIS IDEA. TO DESCRIBE THE PROBABILITY OF A *TYPE II ERROR*, WE BREAK OUT ANOTHER GREEK LETTER: BETA, OR β.

$$\beta = Pr(\text{ACCEPTING } H_0 | H_a)$$

$$= Pr(\text{TYPE II ERROR} | H_a)$$

THE *POWER* OF A TEST IS DEFINED AS $1 - \beta$. IT'S

$$Pr(\text{REJECTING } H_0 | H_a).$$

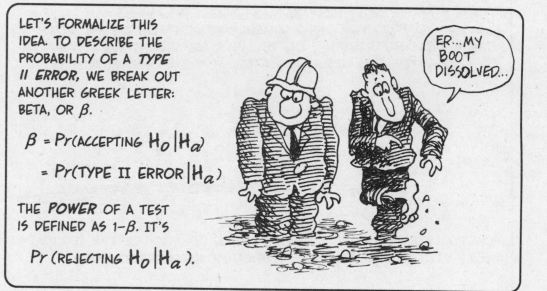

YOU'LL BE HAPPY TO KNOW THE ENVIRONMENTAL REGULATORS HAVE MOVED IN THE DIRECTION OF REQUIRING POLLUTION MONITORING PROGRAMS TO SHOW THAT THEY HAVE A HIGH PROBABILITY OF DETECTING SERIOUS POLLUTION EVENTS. THE REQUIRED *POWER ANALYSIS* OFTEN REVEALS HIDDEN FLAWS IN THE MONITORING PROGRAM.

ONE WAY TO VISUALIZE THE EFFECT OF A TEST'S POWER IS BY GRAPHING THE PROBABILITY OF REJECTING H_0 AGAINST THE ACTUAL STATE OF THE SYSTEM. IN THE CASE OF A SMOKE ALARM, THE PROBABILITY CLIMBS TOWARD 1 AS THE SMOKE GETS THICKER.

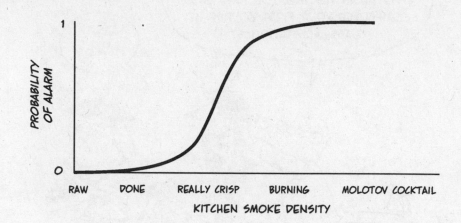

FOR THE E.P.A. WATER QUALITY EXAMPLE, THE HORIZONTAL AXIS IS THE TRUE CONCENTRATION OF POLLUTANT IN THE WATER.

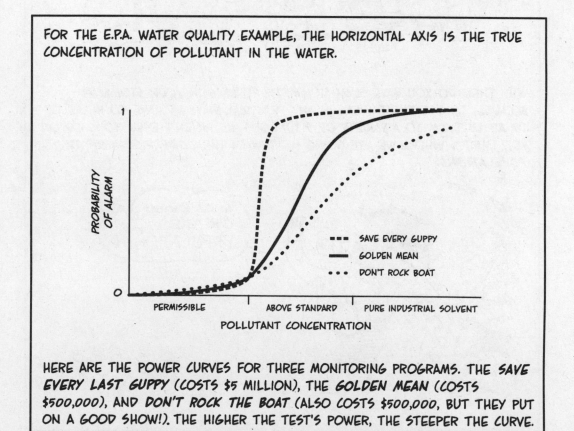

HERE ARE THE POWER CURVES FOR THREE MONITORING PROGRAMS. THE *SAVE EVERY LAST GUPPY* (COSTS $5 MILLION), THE *GOLDEN MEAN* (COSTS $500,000), AND *DON'T ROCK THE BOAT* (ALSO COSTS $500,000, BUT THEY PUT ON A *GOOD SHOW!*). THE HIGHER THE TEST'S POWER, THE STEEPER THE CURVE.

CONGRATULATIONS! WITH THESE SECTIONS COVERING THE BASICS OF CONFIDENCE INTERVALS AND HYPOTHESIS TESTING, YOU HAVE JUST COMPLETED YOUR FIRST COURSE IN CLASSICAL STATISTICS!

WE HAVE?

WHY THEN DO YOU HAVE SUCH AN *EMPTY FEELING* IN YOUR STOMACH? BECAUSE, TO USE THESE IDEAS IN ANY PRACTICAL WAY, WE HAVE TO BE ABLE TO APPLY THEM TO A VARIETY OF SITUATIONS WE HAVEN'T EVEN TOUCHED ON YET. THAT IS WHERE WE ARE GOING NEXT, WITH THE *COMPARISON OF TWO POPULATIONS.*

O.K.! BRING ON THE POPULATIONS!!

◆Chapter 9◆
COMPARING
TWO POPULATIONS

IN WHICH WE LEARN SOME NEW RECIPES USING
OLD INGREDIENTS...

THE LAST TWO CHAPTERS EXPLAINED CONFIDENCE INTERVALS AND HYPOTHESIS TESTING WITH THE **STEAK AND POTATOES** OF RANDOM MODELS: THE **NORMAL** AND THE **BINOMIAL** DISTRIBUTIONS.

WITH THE NORMAL PLAYING THE ROLE OF THE POTATOES!

BUT WHAT MAKES STATISTICS ALMOST AS CHALLENGING AS COOKING IS THE VARIETY. LIKE AN EXPERT COOK, THE STATISTICIAN CAN "TASTE" THE INGREDIENTS IN A PROBLEM AND THEN FIND THE MOST EFFECTIVE WAY TO COMBINE THEM INTO A STATISTICAL RECIPE.

HM... HOW DO YOU SUBTRACT SALT?

(THE REASON COOKBOOKS AND STATISTICAL METHODS TEXTS ARE SO HEAVY IS THAT THEY BOTH PROVIDE SOLUTIONS IN A GREAT VARIETY OF SITUATIONS!)

NOW WHERE IS THAT BINOMIAL SAUCE?

IN THIS CHAPTER, WE'LL USE OUR MEAT AND POTATOES METHODS IN SOME NEW RECIPES THAT WILL HELP US ANSWER THE FOLLOWING QUESTIONS:

MM!

DOES TAKING ASPIRIN REGULARLY REDUCE THE RISK OF HEART ATTACK?

DOES A PARTICULAR PESTICIDE INCREASE THE YIELD OF CORN PER ACRE?

DO MEN AND WOMEN IN THE SAME OCCUPATION HAVE DIFFERENT SALARIES?

THE COMMON INGREDIENT IN THESE QUESTIONS IS THIS: THEY CAN BE ANSWERED BY *COMPARING TWO INDEPENDENT RANDOM SAMPLES*, ONE FROM EACH OF TWO POPULATIONS.

PESTICIDE

NO PESTICIDE

AND, AT THE END OF THE CHAPTER, WE'LL LOOK AT A DIFFERENT WAY TO COMPARE TWO MEANS THAT DOESN'T INVOLVE TAKING TWO SIMPLE RANDOM SAMPLES...

Comparing **SUCCESS RATES**
(or failure rates) for two populations.

WE BEGIN WITH AN EXPERIMENT, PART OF A HARVARD STUDY, THAT SOUGHT TO DECIDE THE EFFECTIVENESS OF *ASPIRIN* IN REDUCING HEART ATTACKS. AS IN MOST CLINICAL TRIALS, THE CHANCES THAT ANY ONE INDIVIDUAL GETS THE DISEASE—IN THIS CASE, A HEART ATTACK—IS VERY SMALL IN ANY GIVEN YEAR. BUT WE WANT ANSWERS QUICKLY! WHAT DO WE DO?

THE SIMPLE, BUT EXPENSIVE, SOLUTION IS TO TEST A LARGE NUMBER OF INDIVIDUALS IN A SHORT TIME. IN THIS STUDY, 22,071 SUBJECTS (ALL VOLUNTEER DOCTORS) WERE RANDOMLY ASSIGNED TO *TWO GROUPS*.

GROUP ONE TOOK A *PLACEBO*—A PILL IDENTICAL TO ASPIRIN, BUT CONTAINING NO ASPIRIN.

GROUP TWO RECEIVED ONE ASPIRIN A DAY.

OVER A PERIOD AVERAGING
NEARLY FIVE YEARS*, THE
INVESTIGATORS RECORDED
THE RESPONSES: HEART
ATTACK OR NO HEART ATTACK.
THE RESULT: (IN THE
NUMBERS THAT FOLLOW, WE
HAVE COMBINED FATAL AND
NONFATAL HEART ATTACKS.)

TSK! SUCCESS WAS FATAL!

	ATTACK	NO ATTACK	n	ATTACK RATE
PLACEBO	239	10,795	11,034	$\hat{p}_1 = \dfrac{239}{11,034} = .0217$
ASPIRIN	139	10,898	11,037	$\hat{p}_2 = \dfrac{139}{11,037} = .0126$

THE OBSERVED DIFFERENCE
IN SUCCESS RATE IS
$\hat{p}_1 - \hat{p}_2 = .0091$. IT SOUNDS
SMALL UNTIL YOU LOOK AT
THE RELATIVE RISK,

$$\frac{\hat{p}_1}{\hat{p}_2} = \frac{.0217}{.0126} = 1.72.$$

MEMBERS OF THE PLACEBO
GROUP WERE 1.72 TIMES
LIKELIER TO SUFFER A HEART
ATTACK THAN THOSE IN THE
ASPIRIN GROUP.

*THE STUDY WAS STOPPED EARLY BECAUSE OF ITS POSITIVE OUTCOME. IT WOULD
HAVE BEEN UNWISE AND IMPRACTICAL TO DENY THE RESULTS TO THE GROUP
TAKING THE PLACEBO.

The Model: THE PLACEBO AND ASPIRIN GROUP OBSERVATIONS ARE INDEPENDENT SAMPLES FROM TWO BINOMIAL POPULATIONS. FOR CONSISTENCY, WE REFER TO A HEART ATTACK AS A *SUCCESS* (!).

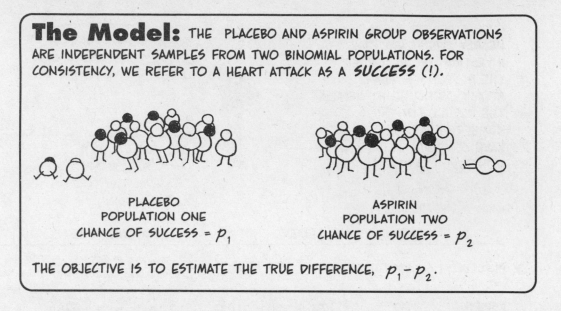

PLACEBO
POPULATION ONE
CHANCE OF SUCCESS = p_1

ASPIRIN
POPULATION TWO
CHANCE OF SUCCESS = p_2

THE OBJECTIVE IS TO ESTIMATE THE TRUE DIFFERENCE, $p_1 - p_2$.

FOR EACH POPULATION (ACTUALLY LARGE SAMPLES OF THE GENERAL POPULATION), WE HAVE THE FAMILIAR RANDOM VARIABLES:

X_1 NUMBER OF SUCCESSES IN POPULATION ONE

X_2 NUMBER OF SUCCESSES IN POPULATION TWO

$\hat{P}_1 = \dfrac{X_1}{n_1}$ PROPORTION OF SUCCESSES IN POPULATION ONE

$\hat{P}_2 = \dfrac{X_2}{n_2}$ PROPORTION OF SUCCESSES IN POPULATION TWO

AND AN ESTIMATOR OF DIFFERENCE IN RATE: $\hat{P}_1 - \hat{P}_2$

AND NOW, LIKE A BROKEN RECORD, WE ASK OURSELVES, HOW IS $\hat{P}_1 - \hat{P}_2$ DISTRIBUTED?

HOW? HOW? HOW?

Sampling distribution for $\hat{P}_1 - \hat{P}_2$

FOR LARGE SAMPLES, $\hat{P}_1 - \hat{P}_2$ IS APPROXIMATELY NORMALLY DISTRIBUTED, MUCH AS IN THE CASE OF ONLY ONE SAMPLE. WE CAN MAKE THE USUAL z-TRANSFORM TO GET A STANDARD NORMAL RANDOM VARIABLE (APPROXIMATELY)

$$z = \frac{\hat{P}_1 - \hat{P}_2 - (p_1 - p_2)}{\sigma(\hat{P}_1 - \hat{P}_2)}$$

BUT HOW DO WE FIND THAT STANDARD DEVIATION IN THE DENOMINATOR?

SINCE THE TWO SAMPLES ARE INDEPENDENT, SO ARE THE RANDOM VARIABLES $\hat{P}_1$ AND $\hat{P}_2$, AND THE TWO VARIANCES ADD.

$$\sigma^2(\hat{P}_1 - \hat{P}_2) = \sigma^2(\hat{P}_1) + \sigma^2(\hat{P}_2)$$

SO

$$\sigma(\hat{P}_1 - \hat{P}_2) = \sqrt{\sigma^2(\hat{P}_1) + \sigma^2(\hat{P}_2)}$$

AND NOW, KNOWING THE DISTRIBUTION OF THE TEST STATISTICS, WE CAN PROCEED TO ESTIMATE *CONFIDENCE INTERVALS* AND *TEST THE HYPOTHESIS* THAT ASPIRIN REDUCES HEART ATTACKS.

Confidence Intervals for $p_1 - p_2$

AS USUAL, THE CONFIDENCE INTERVALS FOR OUR ESTIMATE LOOK LIKE THIS:

$$p_1 - p_2 = \hat{p}_1 - \hat{p}_2 \pm z_{\frac{\alpha}{2}} SE(\hat{p}_1 - \hat{p}_2)$$

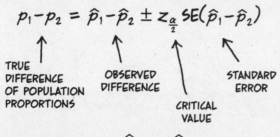

TRUE DIFFERENCE OF POPULATION PROPORTIONS

OBSERVED DIFFERENCE

CRITICAL VALUE

STANDARD ERROR

THE VARIANCES OF $\hat{P}_1$ AND $\hat{P}_2$ ADD, SO THE STANDARD ERROR BECOMES

$$SE(\hat{p}_1 - \hat{p}_2) = \sqrt{\frac{\hat{p}_1(1-\hat{p}_1)}{n_1} + \frac{\hat{p}_2(1-\hat{p}_2)}{n_2}}$$

IN THE ASPIRIN STUDY, THE STANDARD ERROR IS

$$\sqrt{\frac{(.0217)(.9783)}{11,034} + \frac{(.0126)(.9874)}{11,037}}$$

$$= .00175$$

WHAT'S THIS?

STANDARD COOKING ERROR...

TO GET THE 95% CONFIDENCE INTERVAL FOR THE ASPIRIN STUDY, WE JUST PLUG IN THE OBSERVED VALUES:

$$p_1 - p_2 = .0091 \pm (1.96)(.00175)$$

$$= .0091 \pm .0034$$

TRANSLATION:

WE ARE AT LEAST 95% CONFIDENT THAT THE DIFFERENCE IN HEART ATTACK RATES IS BETWEEN .0057 AND .0125. DEFINITELY A POSITIVE NUMBER! WE ARE NOW AT LEAST 95% CONFIDENT THAT ASPIRIN REALLY DOES LOWER HEART ATTACK RATES.

UM...WOULD YOU ADD SOME ASPIRIN TO MY KIBBLE?

hypothesis testing

THE FORMAL HYPOTHESIS-TESTING QUESTION IS

IF ASPIRIN HAD NO EFFECT, WHAT IS THE PROBABILITY THAT THIS RESULT OCCURRED BY CHANCE?

H_0, THE NULL HYPOTHESIS, IS THAT ASPIRIN HAD NO EFFECT: $p_1 = p_2$.

H_a, THE ALTERNATIVE, IS THAT ASPIRIN DOES REDUCE THE HEART ATTACK RATE: $p_1 > p_2$.

NOW WE NEED A TEST STATISTIC WITH A NORMAL DISTRIBUTION WHEN H_0 IS TRUE...

HM! HA! HO!

NOTE THAT UNDER H_0, THE TWO PROPORTIONS ARE THE SAME, $p_1 = p_2 = p$... SO LET'S POOL ALL THE DATA TO GET THE PROPORTION OF HEART ATTACKS IN **BOTH SAMPLES TOGETHER**:

$$\hat{p} = \frac{x_1 + x_2}{n_1 + n_2}$$

WHEN THE NULL HYPOTHESIS IS TRUE, THE STANDARD ERROR DEPENDS ONLY ON THIS POOLED ESTIMATE:

$$SE_0(\hat{P}_1 - \hat{P}_2) = \sqrt{\hat{p}(1-\hat{p})\left(\frac{1}{n_1} + \frac{1}{n_2}\right)}$$

AND WE CAN WRITE A TEST STATISTIC:

$$Z = \frac{\hat{P}_1 - \hat{P}_2}{SE_0(\hat{P}_1 - \hat{P}_2)}$$

(THE NUMERATOR WOULD ORDINARILY BE $\hat{P}_1 - \hat{P}_2 - (p_1 - p_2)$, BUT H_0 ASSUMES $p_1 - p_2 = 0$.)

I HAVE IT! LET'S PLUG INTO THE FORMULA!

EGAD, HOLMES...

FOR THE ASPIRIN STUDY, WE FIND

$$\hat{p} = \frac{378}{22{,}071}$$

$$SE_0(\hat{P}_1 - \hat{P}_2) = .00175$$

SO

$$Z_{OBS} = \frac{.0091}{.00175} = \mathbf{5.20}$$

Z_{OBS} IS MORE THAN *FIVE STANDARD DEVIATIONS* FROM ZERO, A STRONG POSITIVE EFFECT. USING A TABLE OR A COMPUTER, WE FIND THE P-VALUE:

$$\text{P-VALUE} = \text{PR}(Z \geq Z_{OBS}) = \text{PR}(Z \geq 5.2) = .0000001$$

> BY USING A TABLE, A COMPUTER, OR A COMPUTER **ON** A TABLE...

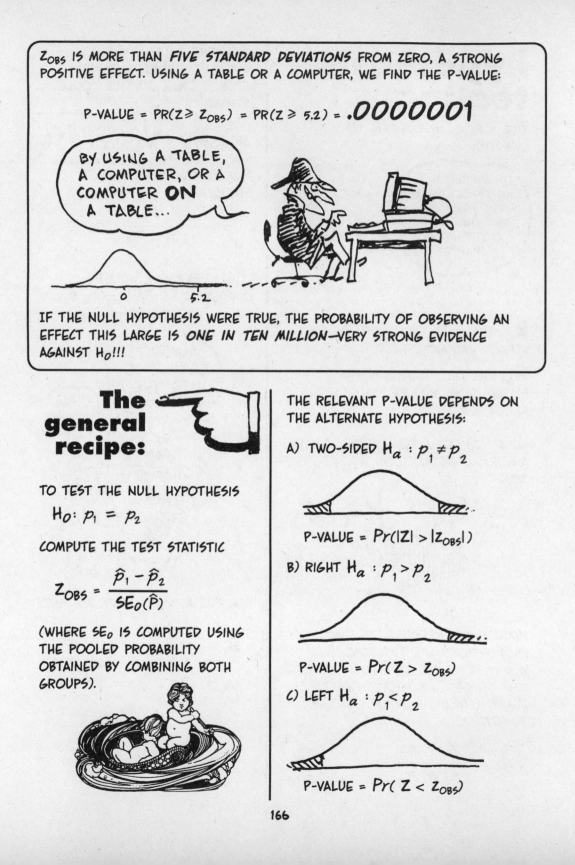

IF THE NULL HYPOTHESIS WERE TRUE, THE PROBABILITY OF OBSERVING AN EFFECT THIS LARGE IS *ONE IN TEN MILLION*—VERY STRONG EVIDENCE AGAINST H_O!!!

The general recipe:

TO TEST THE NULL HYPOTHESIS

$$H_O: p_1 = p_2$$

COMPUTE THE TEST STATISTIC

$$Z_{OBS} = \frac{\hat{p}_1 - \hat{p}_2}{SE_0(\hat{P})}$$

(WHERE SE_0 IS COMPUTED USING THE POOLED PROBABILITY OBTAINED BY COMBINING BOTH GROUPS).

THE RELEVANT P-VALUE DEPENDS ON THE ALTERNATE HYPOTHESIS:

A) TWO-SIDED $H_a : p_1 \neq p_2$

$$\text{P-VALUE} = Pr(|Z| > |Z_{OBS}|)$$

B) RIGHT $H_a : p_1 > p_2$

$$\text{P-VALUE} = Pr(Z > Z_{OBS})$$

C) LEFT $H_a : p_1 < p_2$

$$\text{P-VALUE} = Pr(Z < Z_{OBS})$$

THE ANALYSIS OF THE ASPIRIN STUDY DEPENDED ON CERTAIN FEATURES OF THE EXPERIMENT DESIGNED TO ENSURE RANDOMNESS AND TO ELIMINATE BIAS:

1 SUBJECTS WERE RANDOMLY ASSIGNED TO TREATMENT GROUPS.

2 THE EXPERIMENT WAS *BLIND:* SUBJECTS DIDN'T KNOW IF THEY WERE TAKING ASPIRIN OR PLACEBO.

3 THE SAMPLE SIZE WAS LARGE ENOUGH FOR THE NORMAL APPROXIMATION TO WORK.

$np > 5$, RIGHT?

RIGHT...

POINTS 1 AND 2 ARE ESSENTIAL PARTS OF MOST HUMAN CLINICAL TRIAL DESIGNS, BUT POINT 3 IS NOT ESSENTIAL. GOOD SMALL-SAMPLE TESTS DO EXIST AND ARE AVAILABLE IN COMPUTER SOFTWARE PACKAGES. THESE *NONPARAMETRIC* PROCEDURES DEPEND ON SIMPLE, BUT LENGTHY, PROBABILITY CALCULATIONS SIMILAR TO THE GAMBLING COMPUTATIONS WE ENCOUNTERED IN CHAPTER 4...

UM... WE ALSO ASSUMED THAT DOCTORS ARE REPRESENTATIVE OF THE GENERAL POPULATION...

HOW INSULTING.

Comparing the MEANS of two populations

SUPPOSE WE WANTED TO COMPARE THE
AVERAGE SALARY OF MALE AND FEMALE
EMPLOYEES IN THE SAME JOB AT SOME
COMPANY.

POPULATION ONE IS THE WOMEN, AND POPULATION TWO IS THE MEN.

POPULATION ONE HAS MEAN
SALARY μ_1 AND STANDARD
DEVIATION σ_1

POPULATION TWO HAS MEAN
SALARY μ_2 AND STANDARD
DEVIATION σ_2

A RANDOM SAMPLE OF SIZE n_1 FROM GROUP 1 AND n_2 FROM GROUP 2 GIVES
SAMPLE MEANS OF $\bar{x}_1$ AND $\bar{x}_2$ AND STANDARD DEVIATIONS S_1 AND S_1,
RESPECTIVELY. THE ESTIMATOR OF $\mu_1 - \mu_2$ IS

$$\bar{X}_1 - \bar{X}_2$$

HOW GOOD AN ESTIMATOR IS $\bar{X}_1 - \bar{X}_2$? FOR LARGE SAMPLE SIZES, IT'S APPROXIMATELY NORMAL (BY THE CENTRAL LIMIT THEOREM), AND ITS STANDARD ERROR IS

$$SE(\bar{X}_1 - \bar{X}_2) = \sqrt{\frac{s_1^2}{n_1} + \frac{s_2^2}{n_2}}$$

(THE VARIANCES ADD, SINCE SAMPLES ARE INDEPENDENT.) NOW WE CAN PROCEED DIRECTLY TO:

confidence intervals: FOR

LARGE SAMPLE SIZES, THE $(1-\alpha)100\%$ CONFIDENCE INTERVAL FOR THE DIFFERENCE BETWEEN MEANS IS

$$\mu_1 - \mu_2 = \bar{x}_1 - \bar{x}_2 \pm z_{\frac{\alpha}{2}} SE(\bar{X}_1 - \bar{X}_2)$$

HEY, GUYS! LOOK! $SE(\bar{X})$! RIGHT IN THE FORMULA!!

WHAT A STUPID JOKE...

hypothesis testing: WE ASSESS

THE NULL HYPOTHESIS THAT THE TWO POPULATION MEANS ARE EQUAL.

$$H_0: \quad \mu_1 = \mu_2$$

THE TEST STATISTIC IS

$$Z_{OBS} = \frac{\bar{X}_1 - \bar{X}_2}{SE(\bar{X}_1 - \bar{X}_2)}$$

AND THE P-VALUES WORK IN THE USUAL WAY.

P-VALUE! DID YOU HEAR THAT? P-VALUE!

and how about comparing
SMALL SAMPLE
MEANS?

REMEMBER *CHAMELEON MOTORS*? THEIR COMPETITOR, *IGUANA AUTO*, CLAIMS THAT ITS STYROFOAM HOOD ORNAMENT GIVES BETTER FRONT END CRASH PROTECTION, AND THEY'VE CRASHED *SEVEN IGUANAS* TO PROVE IT!

THEIR RESULTS, COMPARED WITH CHAMELEON'S:

CHAMELEON	
1	$150
2	$400
3	$720
4	$500
5	$930
n_1	5
$\bar{x}_1$	$540
s_1	$299

IGUANA	
1	$50
2	$200
3	$150
4	$400
5	$750
6	$400
7	$150
n_2	7
$\bar{x}_2$	$300
s_2	$238

THE t DISTRIBUTION CAN BE USED IF BOTH POPULATIONS ARE MOUND SHAPED AND HAVE THE SAME STANDARD DEVIATION $\sigma = \sigma_1 = \sigma_2$. THE ONLY WRINKLE IS THAT WE HAVE TO *POOL* THE SUM OF SQUARES ABOUT THE MEANS TO FORM A SINGLE ESTIMATE OF σ:

$$S^2_{POOL} = \frac{(n_1-1)s_1^2 + (n_2-1)s_2^2}{n_1 + n_2 - 2}$$

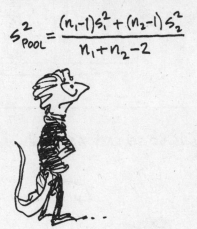

- THE STANDARD ERROR IS THE SAME AS FOR LARGE SAMPLES, EXCEPT THAT S_{POOL} REPLACES S_1 AND S_2:

$$SE(\overline{X}_1 - \overline{X}_2) = \sqrt{\frac{S^2_{POOL}}{n_1} + \frac{S^2_{POOL}}{n_2^2}}$$

$$= S_{POOL}\sqrt{\frac{1}{n_1} + \frac{1}{n_2}}$$

THE $(1-\alpha)\cdot100\%$ CONFIDENCE INTERVAL IS

$$\mu_1 - \mu_2 = \overline{x}_1 - \overline{x}_2 \pm t_{\frac{\alpha}{2}} SE(\overline{X}_1 - \overline{X}_2)$$

WHERE $t_{\frac{\alpha}{2}}$ IS A CRITICAL VALUE OF t WITH $n_1 - n_2 - 2$ DEGREES OF FREEDOM.

THE REPTILIAN CARMAKERS AGREE THAT THEIR STANDARD DEVIATIONS ARE CLOSE AND THEIR REPAIR HISTOGRAMS ARE MOUND-SHAPED. THEY COMPUTE:

$$S_{POOL} = \sqrt{\frac{4\cdot299^2 + 6\cdot328^2}{10}} = 264$$

$$SE(\overline{X}_1 - \overline{X}_2) = 264\sqrt{\frac{1}{5} + \frac{1}{7}} = 154$$

THE 95% CONFIDENCE INTERVAL IS

$$\mu_1 - \mu_2 = 540 - 300 \pm t_{.025}(154)$$
$$= 240 \pm (2.23)(154)$$
$$= \mathbf{240 \pm 340}$$

SINCE THIS INCLUDES THE VALUE 0, IGUANA AUTOS HAS *NOT* SHOWN A SIGNIFICANT IMPROVEMENT IN REPAIR COSTS.

O.K...FORGET SAFETY... BUT YOU CAN'T ARGUE WITH BEAUTIFUL STYLING...

HERE'S AN EXAMPLE THAT SHOWS THE PITFALLS OF MINDLESSLY FOLLOWING THE COOKBOOK: A LARGE TAXI FLEET OWNER WANTS TO COMPARE THE GAS MILEAGE USING GAS A AND GAS B.

STARTING WITH 100 CABS, HE RANDOMLY ASSIGNS 50 TO EACH GASOLINE, AND, AFTER A DAY'S DRIVING, DETERMINES

	SAMPLE SIZE	MEAN MILEAGE	STANDARD DEVIATION
A	50	25	5.00
B	50	26	4.00

THE SAMPLE DIFFERENCE IS

$$\bar{x}_1 - \bar{x}_2 = 25 - 26 = -1$$

IS GAS B REALLY BETTER THAN GAS A?

OWING TO THE LARGE STANDARD DEVIATIONS, THE STANDARD ERROR IS PRETTY SUBSTANTIAL:

$$SE(\bar{X}_1 - \bar{X}_2) = \sqrt{\frac{s_1^2}{n_1} + \frac{s_2^2}{n_2}}$$

$$= \sqrt{\frac{25}{50} + \frac{16}{50}}$$

$$= .905$$

AT THE 95% CONFIDENCE LEVEL, WE HAVE

$$\mu_1 - \mu_2 = \bar{x}_1 - \bar{x}_2 \pm z_{.025}(.905)$$

$$= -1 \pm (1.96)(.905)$$

$$= -1 \pm 1.774$$

THIS INCLUDES THE VALUE 0, CORRESPONDING TO $\mu_1 = \mu_2$

AND HYPOTHESIS TESTING?

THE P-VALUE FOR THE ALTERNATE HYPOTHESIS, H_a: $\mu_1 \neq \mu_2$, IS

$$Pr(|z| \geq |z_{OBS}|) = Pr\left(|z| \geq \frac{1}{.905}\right)$$

$$= Pr(|z| \geq 1.1) = 2(.1357)$$

$$= .2714$$

TOTAL SHADED AREA = .2714

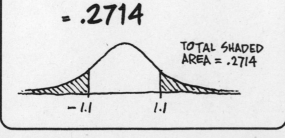

THIS **EXCEEDS** THE $\alpha = .05$ SIGNIFICANCE LEVEL, SO WE CONCLUDE THAT THE EVIDENCE IN FAVOR OF EITHER GAS IS VERY WEAK.

CALL ME A STATISTICIAN!!

YOU'RE A STATISTICIAN!

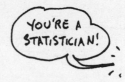

PAIRED COMPARISONS
a better way to compare gasolines

THE TAXI OWNER FOLLOWED THE COOKBOOK EXACTLY. HIS SAMPLES WERE RANDOM, AND HIS SAMPLE SIZE WAS LARGE ENOUGH. HE JUST FAILED TO *THINK* WHEN NECESSARY!

ALTHOUGH GAS B APPEARS TO BE SLIGHTLY BETTER THAN GAS A, THE CONFIDENCE INTERVAL WAS WIDE BECAUSE OF THE LARGE STANDARD DEVIATIONS—I.E., THE *MILEAGES VARIED WIDELY FROM ONE CAB TO THE NEXT.* WHY SUCH HIGH VARIABILITY? BECAUSE CABS—AND CABBIES—HAVE DIFFERENT *PERSONALITIES!*

A FAR BETTER WAY TO DO THIS STUDY IS TO ASSIGN GAS A AND GAS B TO THE *SAME CAB ON DIFFERENT DAYS.*

WE STILL RANDOMIZE THE TREATMENT BY FLIPPING A COIN TO DECIDE WHETHER TO USE GAS A ON TUESDAY OR WEDNESDAY. WE CAN ALSO CUT THE EXPERIMENT DOWN TO 10 CABS, SAVING THE OWNER A LOT OF TIME AND MONEY!

WAY FEWER COINS TO TOSS!

CAB	GAS A	GAS B	DIFFERENCE
1	27.01	26.95	0.06
2	20.00	20.44	− 0.44
3	23.41	25.05	− 1.64
4	25.22	26.32	− 1.10
5	30.11	29.56	0.55
6	25.55	26.60	− 1.05
7	22.23	22.93	− 0.70
8	19.78	20.23	− 0.45
9	33.45	33.95	− 0.50
10	25.22	26.01	− 0.79
MEAN	25.20	25.80	− 0.60
STANDARD DEVIATION	4.27	4.10	0.61

NOTE THAT THE MEANS AND STANDARD DEVIATIONS OF GAS A AND GAS B ARE ABOUT THE SAME. THAT'S TO BE EXPECTED, SINCE THEY HAVE THE SAME SOURCE OF VARIABILITY AS IN THE UNPAIRED EXPERIMENT. BUT NOW THE *DIFFERENCE COLUMN* HAS A VERY *SMALL* STANDARD DEVIATION. THE DIFFERENCE COLUMN, BY COMPARING GAS PERFORMANCE WITHIN A SINGLE CAR, ELIMINATES VARIABILITY BETWEEN TAXIS.

THE DIFFERENCES d_i PROVIDE A **SINGLE MEASURE OF DIFFERENCE FOR EACH TAXI**, AND WE CAN USE IT TO MAKE A SMALL-SAMPLE t TEST STATISTIC:

$$t = \frac{\bar{d}}{s_d/\sqrt{n}}$$

THE 95% CONFIDENCE INTERVAL AROUND $\bar{d}$ IS

$$\mu_d = \bar{d} \pm t_{.025}\left(s_d/\sqrt{n}\right)$$

SAMPLE MEAN   CRITICAL VALUE   STANDARD ERROR

$$= -.6 \pm (2.26)\left(\frac{.61}{\sqrt{10}}\right)$$
$$= -.60 \pm .44$$

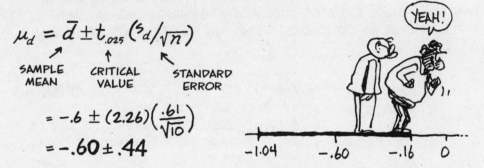

SO WE HAVE $-1.04 \leq \mu_d \leq -.16$ WITH 95% CONFIDENCE, GOOD EVIDENCE THAT GAS B REALLY IS BETTER.

THE HYPOTHESIS-TESTING P-VALUE CAN BE FOUND USING A SOFTWARE PACKAGE:

$$H_a: \mu_d \neq 0$$
$$\text{P-VALUE} = Pr\left(|t| \geq |t_{OBS}|\right)$$
$$= Pr\left(|t| \geq \frac{.6}{.19}\right)$$
$$= Pr\left(|t| \geq 3.15\right)$$
$$= .012 < .05$$

AGAIN, GAS B PASSES THE TEST.

HERE ARE PLOTS OF THE GAS MILEAGE DATA: THE FIRST ONE SHOWS THE MILEAGES UNPAIRED:

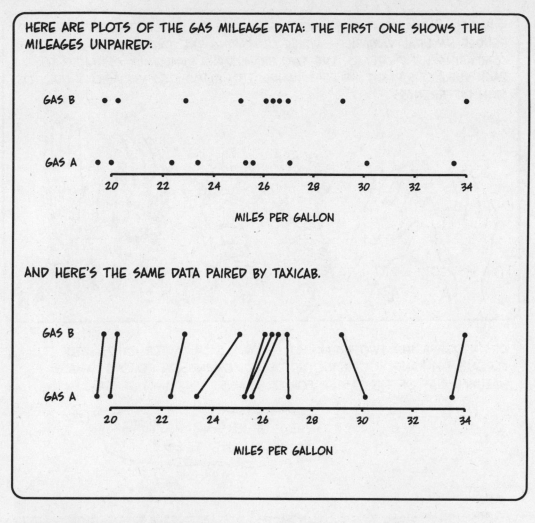

AND HERE'S THE SAME DATA PAIRED BY TAXICAB.

THE PREDOMINANCE OF RIGHT-LEANING LINES IS A STRONG HINT THAT GAS B GIVES BETTER MILEAGE.

WHAT RIGHT-LEANING LINES?

A PAIRED COMPARISON EXPERIMENT IS ONE OF THE MOST EFFECTIVE WAYS TO REDUCE NATURAL VARIABILITY WHILE COMPARING TREATMENTS. FOR EXAMPLE, IN COMPARING HAND CREAMS, THE TWO BRANDS ARE RANDOMLY ASSIGNED TO EACH SUBJECT'S RIGHT OR LEFT HANDS. THIS ELIMINATES VARIABILITY DUE TO SKIN DIFFERENCES.

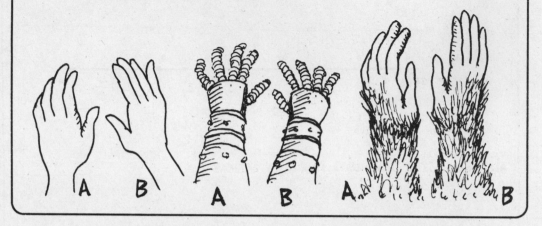

OR, IN COMPARING TWO BREAKFAST CEREALS, EACH TASTER RATES BOTH CEREALS (IN RANDOM ORDER). THE PAIRED COMPARISON REMOVES THE NATURAL BIAS OF THE TASTER FOR OR AGAINST CEREAL IN GENERAL.

IN THIS CHAPTER, WE APPLIED THE BASIC IDEAS ABOUT CONFIDENCE INTERVALS AND HYPOTHESIS TESTING TO THE COMPARISON OF TWO POPULATIONS. THERE ARE INNUMERABLE FURTHER POSSIBILITIES. WE COULD HAVE GONE ON TO DESCRIBE COMPARISONS OF:

- THE STANDARD DEVIATIONS OF TWO POPULATIONS WHEN SAMPLE SIZE IS SMALL,

- THE MEANS OF MORE THAN TWO POPULATIONS WHEN SAMPLE SIZE IS LARGE,

- THE MEANS OF MORE THAN TWO POPULATIONS WHEN SAMPLE SIZE IS SMALL,

ETC!

THIS IS WHY STATISTICS BOOKS ARE SO THICK...

IN PRACTICE, STATISTICIANS DETERMINE THE GENERAL NATURE OF THE PROBLEM, AND THEN CONSULT THE RIGHT REFERENCE BOOK.

THE ONLY THING REALLY NEW IN THE CHAPTER WAS THE IDEA OF THE *PAIRED COMPARISON TEST.* IN THE NEXT CHAPTER, WE'LL LOOK AT SOME OTHER KINDS OF EXPERIMENTAL DESIGNS.

♦Chapter 10♦

EXPERIMENTAL DESIGN

THE DESIGN OF AN EXPERIMENT OFTEN SPELLS SUCCESS OR FAILURE. IN THE PAIRED COMPARISONS EXAMPLE, OUR STATISTICIAN CHANGED ROLES FROM PASSIVE NUMBER GATHERING AND ANALYSIS TO ACTIVE PARTICIPATION IN THE DESIGN OF THE EXPERIMENT.

IN THIS CHAPTER, WE INTRODUCE THE BASIC IDEAS OF EXPERIMENTAL DESIGN, WHILE LEAVING THE DETAILED NUMERICAL ANALYSIS TO YOUR HANDY STATISTICAL SOFTWARE PACK.

NO FORMULAS IN THIS CHAPTER... SORRY!

THE ELEMENTS OF A DESIGN ARE THE *EXPERIMENTAL UNITS* AND THE *TREATMENTS* THAT ARE TO BE ASSIGNED TO THE UNITS. THE OBJECTIVE OF ANY DESIGN IS TO COMPARE THE TREATMENTS.

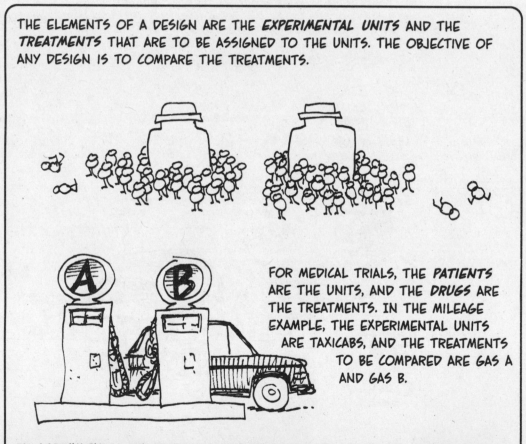

FOR MEDICAL TRIALS, THE *PATIENTS* ARE THE UNITS, AND THE *DRUGS* ARE THE TREATMENTS. IN THE MILEAGE EXAMPLE, THE EXPERIMENTAL UNITS ARE TAXICABS, AND THE TREATMENTS TO BE COMPARED ARE GAS A AND GAS B.

IN AGRICULTURAL EXPERIMENTS, THE EXPERIMENTAL UNITS ARE OFTEN PLOTS IN A FIELD, AND THE TREATMENTS MIGHT BE APPLICATION OF DIFFERENT WHEAT VARIETIES, PESTICIDES, FERTILIZERS, ETC.

TODAY, EXPERIMENTAL DESIGN IDEAS ARE USED EXTENSIVELY IN *INDUSTRIAL PROCESS OPTIMIZATION, MEDICINE* AND *SOCIAL SCIENCE.* ANY EXPERIMENTAL DESIGN USES *THREE BASIC PRINCIPLES,* WHICH ARE CLEARLY ILLUSTRATED IN OUR CAB EXAMPLE:

I ALWAYS KNEW THAT DRIVING A CAB WAS A FORM OF SOCIAL SCIENCE...

Replication: THE SAME TREATMENTS ARE ASSIGNED TO DIFFERENT EXPERIMENTAL UNITS. WITHOUT REPLICATION, IT'S IMPOSSIBLE TO ASSESS NATURAL VARIABILITY AND MEASUREMENT ERROR.

Local control REFERS TO ANY METHOD THAT ACCOUNTS FOR AND REDUCES NATURAL VARIABILITY. ONE WAY IS TO GROUP SIMILAR EXPERIMENTAL UNITS INTO *BLOCKS.* IN THE CAB EXAMPLE, BOTH GASOLINES WERE USED IN EACH CAR, AND WE SAY THAT THE CAB IS A BLOCK.

TAKE ME AROUND THE BLOCK!

LADY, YOU'RE **IN** THE BLOCK...

Randomization: THE ESSENTIAL STEP IN ALL STATISTICS! TREATMENTS MUST BE ASSIGNED RANDOMLY TO EXPERIMENTAL UNITS. FOR EACH TAXI, WE ASSIGNED GAS A TO TUESDAY OR WEDNESDAY BY FLIPPING A COIN. IF WE HADN'T, THE RESULTS COULD HAVE BEEN RUINED BY DIFFERENCES BETWEEN TUESDAY AND WEDNESDAY!

TUES

WED

NOW SUPPOSE WE WANT TO INVESTIGATE THE EFFECT OF TWO BRANDS OF TIRES AS WELL AS TWO GASOLINES. WE HAVE FOUR POSSIBLE TREATMENTS, WHICH WE CAN LAY OUT IN A TWO-BY-TWO FACTORIAL DESIGN. THE TWO FACTORS ARE GAS AND TIRE MAKE.

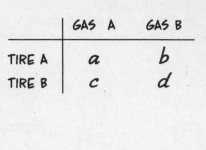

	GAS A	GAS B
TIRE A	a	b
TIRE B	c	d

WE CAN ASSIGN THE FOUR TREATMENTS AT RANDOM TO FOUR DIFFERENT DAYS FOR EACH CAB. ALL FOUR TREATMENTS (a, b, c, AND d) ARE REPEATED WITHIN EACH BLOCK (CAB). THIS IS CALLED A **COMPLETE RANDOMIZED BLOCK DESIGN.**

SO FAR, WE HAVE ASSUMED THAT EVERY DAY OF THE WEEK IS THE SAME, BUT WE CAN CONTROL FOR THIS, TOO, IN THE FOLLOWING WAY: USE ONLY **FOUR CABS,** AND ASSIGN THE TREATMENT ACCORDING TO THE TABLE AT RIGHT:

		DAY			
		1	2	3	4
CAB	1	a	b	c	d
	2	b	c	d	a
	3	c	d	a	b
	4	d	a	b	c

NOTE: EACH TREATMENT APPEARS ONCE IN EACH ROW AND COLUMN!

A FOUR-BY-FOUR TABLE
WITH FOUR DIFFERENT
ELEMENTS, EACH APPEARING
ONCE IN EVERY COLUMN
AND ROW, IS CALLED A
Latin square.
IN THIS EXPERIMENT, THE
FOUR DAYS AND FOUR CABS
GET ALL FOUR TREATMENTS
EXACTLY ONCE.

THE RANDOMIZATION STEP
PICKS A SINGLE LATIN SQUARE
DESIGN AT RANDOM FROM A
LIST OF ALL POSSIBLE FOUR-
WAY LATIN SQUARES.

IF FOUR UNITS ISN'T ENOUGH, WE CAN INCREASE THE NUMBER OF
EXPERIMENTAL UNITS BY *REPEATING* THE EXPERIMENTAL DESIGN. STARTING
WITH EIGHT CABS, WE COULD DIVIDE THEM INTO TWO GROUPS OF FOUR AND
THEN REPEAT THE DESIGN WITHIN EACH GROUP.

O.K... CAR 6 GOES
WITH GAS B AND
TIRE A ON DAY 2 ...
:WHEW:

WE PROMISED NOT TO GO INTO THE DATA ANALYSIS IN ANY DETAIL, BUT HERE IS ROUGHLY HOW A COMPLEX DESIGN LIKE THIS IS HANDLED.

EXPERIMENTAL DESIGNS ARE ANALYZED BY ALLOCATING TOTAL VARIABILITY AMONG DIFFERENT SOURCES. IN THE CAB EXAMPLE, THE SOURCES OF VARIABILITY ARE THE CAB, THE TIRE MAKE, GAS TYPE, DAY—AND RANDOM ERROR. ANALYSIS OF VARIANCE, *ANOVA* FOR SHORT, PARTITIONS THE TOTAL VARIATION, ALLOCATING PORTIONS TO EACH SOURCE.

IN THE NEXT CHAPTER, WE EXPLAIN IN DETAIL ONE MODEL FOR ANALYZING COMPLEX DESIGNS: THE *LINEAR REGRESSION MODEL*. IN LINEAR REGRESSION, YOU'LL BE ABLE TO SEE *ANOVA* UP CLOSE AND NUMERICAL...

◆Chapter 11◆
REGRESSION

SO FAR, WE'VE DONE STATISTICS ON *ONE VARIABLE AT A TIME*, WHETHER IT CAME FROM A POPULATION OF PILL TAKERS, PICKLES, OR CRASHED CARS. IN THIS CHAPTER, WE'LL SEE HOW TO RELATE *TWO VARIABLES*: GIVEN THE *WEIGHTS* OF THE 92 STUDENTS IN CHAPTER 2, WE ASK HOW THEY ARE RELATED TO THE STUDENTS' *HEIGHTS*.

ALL THE BIG QUESTIONS ARE ABOUT RELATIONSHIPS!

THIS IS AN EXAMPLE OF A BROAD CLASS OF IMPORTANT QUESTIONS: DOES *BLOOD PRESSURE LEVEL* PREDICT *LIFE EXPECTANCY?* DO *S.A.T. SCORES* PREDICT *COLLEGE PERFORMANCE?* DOES READING STATISTICS BOOKS MAKE YOU A *BETTER PERSON?*

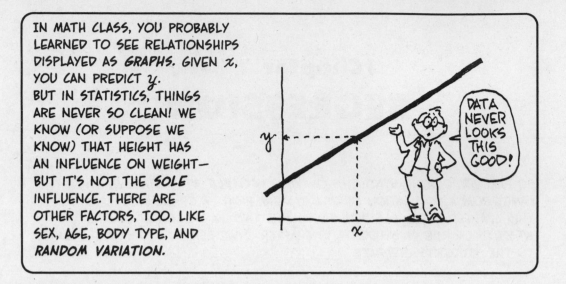

IN MATH CLASS, YOU PROBABLY LEARNED TO SEE RELATIONSHIPS DISPLAYED AS *GRAPHS*. GIVEN x, YOU CAN PREDICT y. BUT IN STATISTICS, THINGS ARE NEVER SO CLEAN! WE KNOW (OR SUPPOSE WE KNOW) THAT HEIGHT HAS AN INFLUENCE ON WEIGHT—BUT IT'S NOT THE *SOLE* INFLUENCE. THERE ARE OTHER FACTORS, TOO, LIKE SEX, AGE, BODY TYPE, AND *RANDOM VARIATION*.

DATA NEVER LOOKS THIS GOOD!

FOR THIS CHAPTER, LET'S LABEL THE WEIGHT DATA AS y AND THE HEIGHT DATA AS x. THUS (x_i, y_i) IS THE HEIGHT AND WEIGHT OF STUDENT i. WE DISPLAY THE POINTS (x_i, y_i) IN A 2-DIMENSIONAL DOT PLOT CALLED A *SCATTERPLOT*.

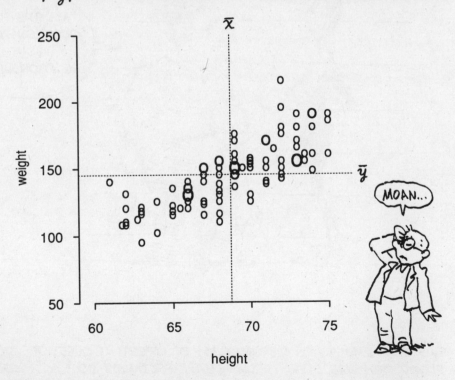

MOAN...

(SOME OF THE DOTS ARE BIGGER, BECAUSE THEY REPRESENT TWO OR THREE STUDENTS WITH THE SAME HEIGHT AND WEIGHT.)

CAN WE PREDICT A STUDENT'S WEIGHT y FROM HIS OR HER HEIGHT x?

Regression analysis

FITS A STRAIGHT LINE TO
THIS MESSY SCATTERPLOT.
x IS CALLED THE
INDEPENDENT OR
PREDICTOR VARIABLE, AND
y IS THE **DEPENDENT** OR
RESPONSE VARIABLE. THE
REGRESSION OR **PREDICTION**
LINE HAS THE FORM

$$y = a + bx$$

TO ILLUSTRATE THE FITTING PROCESS, LET'S USE A SMALLER, RIGGED DATA SET
WITH ONLY **NINE** STUDENT HEIGHT-WEIGHT PAIRS:

HEIGHT	WEIGHT
60	84
62	95
64	140
66	155
68	119
70	175
72	145
74	197
76	150

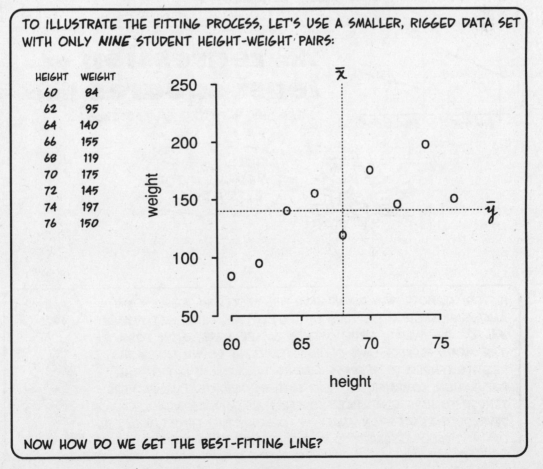

NOW HOW DO WE GET THE BEST-FITTING LINE?

THE IDEA IS TO *MINIMIZE* THE TOTAL SPREAD OF THE y VALUES FROM THE LINE. JUST AS WHEN WE DEFINED THE VARIANCE, WE LOOK AT ALL THE *SQUARED y DISTANCES* FROM THE LINE, AND ADD THEM UP TO GET THE *SUM OF SQUARED ERRORS:*

$$SSE = \sum_{i=1}^{n} (y_i - \hat{y}_i)^2$$

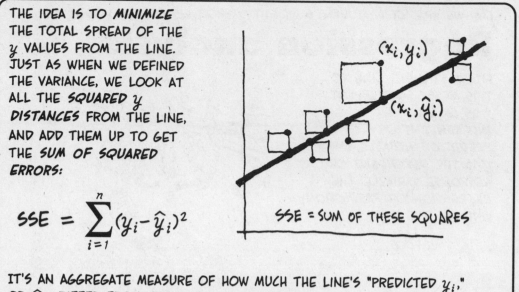

SSE = SUM OF THESE SQUARES

IT'S AN AGGREGATE MEASURE OF HOW MUCH THE LINE'S "PREDICTED y_i," OR $\hat{y}_i$, DIFFER FROM THE ACTUAL DATA VALUES y_i.

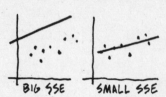

BIG SSE   SMALL SSE

The regression or least squares line
IS THE LINE WITH THE *SMALLEST SSE.*

SHALL WE JUST MEASURE IT FOR EVERY LINE?

HISTORICAL NOTE: WHY DO WE CALL THIS PROCEDURE *REGRESSION ANALYSIS?* AROUND THE TURN OF THE CENTURY, GENETICIST *FRANCIS GALTON* DISCOVERED A PHENOMENON CALLED *REGRESSION TOWARD THE MEAN.* SEEKING LAWS OF INHERITANCE, HE FOUND THAT SONS' HEIGHTS TENDED TO *REGRESS* TOWARD THE MEAN HEIGHT OF THE POPULATION, COMPARED TO THEIR FATHERS' HEIGHTS. TALL FATHERS TENDED TO HAVE SOMEWHAT SHORTER SONS, AND VICE VERSA. GALTON DEVELOPED *REGRESSION ANALYSIS* TO STUDY THIS EFFECT, WHICH HE OPTIMISTICALLY REFERRED TO AS "REGRESSION TOWARD MEDIOCRITY."

GROW UP, BOY!

NOT TO BEAT AROUND THE BUSH, WE GIVE WITHOUT PROOF THE REGRESSION LINE'S FORMULA: IT'S MESSY BUT COMPUTABLE.

$$y = a + bx$$

WHERE

$$b = \frac{\sum\limits_{i=1}^{n}(x_i - \bar{x})(y_i - \bar{y})}{\sum\limits_{i=1}^{n}(x_i - \bar{x})^2}$$

AND

$$a = \bar{y} - b\bar{x}$$

(HERE $\bar{x}$ AND $\bar{y}$ ARE THE MEANS OF $\{x_i\}$ AND $\{y_i\}$ RESPECTIVELY.)

BECAUSE SOME OF THESE EXPRESSIONS WILL SHOW UP AGAIN, WE ABBREVIATE THEM:

$$ss_{xx} = \sum\limits_{i=1}^{n}(x_i - \bar{x})^2$$

$$ss_{yy} = \sum\limits_{i=1}^{n}(y_i - \bar{y})^2$$

SUM OF SQUARES AROUND THE MEAN, THESE MEASURE THE SPREAD OF x_i AND y_i.

$$ss_{xy} = \sum\limits_{i=1}^{n}(x_i - \bar{x})(y_i - \bar{y})$$

THE CROSS PRODUCT DETERMINES (WITH ss_{xx}) THE COEFFICIENT b.

191

FOR THE RIGGED DATA, HERE'S THE WHOLE COMPUTATION:

x_i	y_i	$(x_i - \overline{x})$	$(y_i - \overline{y})$	$(x_i - \overline{x})^2$	$(y_i - \overline{y})^2$	$(x_i - \overline{x})(y_i - \overline{y})$
60	84	−8	−56	64	3136	448
62	95	−6	−45	36	2025	270
64	140	−4	0	16	0	0
66	155	−2	15	4	225	−30
68	119	0	−21	0	441	0
70	175	2	35	4	1225	70
72	145	4	5	16	25	20
74	197	6	57	36	3249	342
76	150	8	10	64	100	80

SUM=612   1260 SS_{xx} =240 SS_{yy} = 10426 SS_{xy} =1200

$\overline{x}$=68 $\overline{y}$=140

WHICH GIVES VALUES OF a AND b:

$$b = \frac{1200}{240} = 5 \qquad a = \overline{y} - b\overline{x} = 140 - 5(68) = -200$$

SO $y = -200 + 5x$

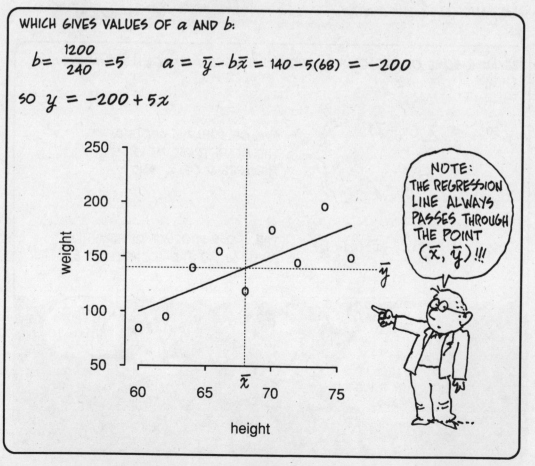

NOTE:
THE REGRESSION
LINE ALWAYS
PASSES THROUGH
THE POINT
$(\overline{x}, \overline{y})$!!!

ANOVA

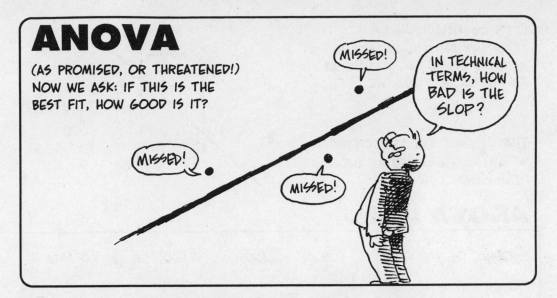

(AS PROMISED, OR THREATENED!)
NOW WE ASK: IF THIS IS THE
BEST FIT, HOW GOOD IS IT?

MISSED!

MISSED!

MISSED!

IN TECHNICAL TERMS, HOW BAD IS THE SLOP?

AS YOU CAN IMAGINE, THE ANSWER TO THIS QUESTION DEPENDS ON HOW
SLOPPILY THE DATA POINTS ARE SPREAD OUT, I.E., HOW BIG **SSE** IS, RELATIVE
TO THE TOTAL SPREAD OF THE DATA. SOME EXAMPLES:

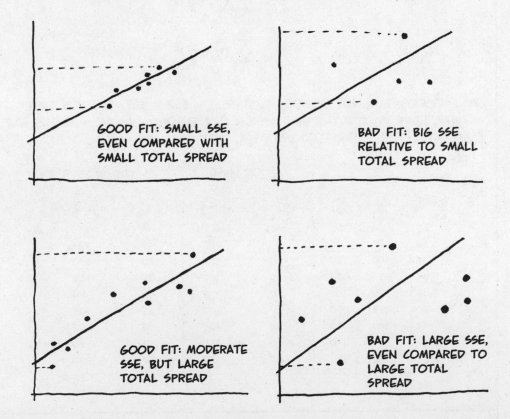

GOOD FIT: SMALL SSE,
EVEN COMPARED WITH
SMALL TOTAL SPREAD

BAD FIT: BIG SSE
RELATIVE TO SMALL
TOTAL SPREAD

GOOD FIT: MODERATE
SSE, BUT LARGE
TOTAL SPREAD

BAD FIT: LARGE SSE,
EVEN COMPARED TO
LARGE TOTAL
SPREAD

LET'S QUANTIFY THIS BY
APPORTIONING THE VARIABILITY
IN y. REFER TO THE PICTURE AT
RIGHT FOR GUIDANCE. WE LET

$$\widehat{y}_i = a + bx_i$$

THUS, $\widehat{y}_i$ ARE THE PREDICTED
WEIGHTS DETERMINED BY THE
REGRESSION LINE.

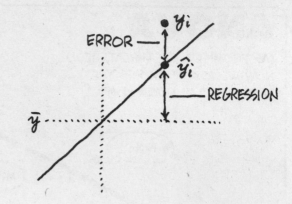

ANOVA table

SOURCE OF VARIABILITY	SUM OF SQUARES	VALUE FOR RIGGED DATA
REGRESSION	$SSR = \sum\limits_{i=1}^{n}(\widehat{y}_i - \overline{y})^2$	6000
ERROR	$SSE = \sum\limits_{i=1}^{n}(y_i - \widehat{y}_i)^2$	4426
TOTAL	$SS_{yy} = \sum\limits_{i=1}^{n}(y_i - \overline{y})^2$	10,426

(BY THE WAY, IT IS NOT OBVIOUS THAT SS_{yy} = SSR + SSE—BUT IT'S TRUE!)
ANYWAY, HERE IS HOW THE REGRESSION AND ERROR SUMS OF SQUARES ARE
CALCULATED FOR THE RIGGED DATA SET, WITH $y = -200 + 5x$.

			REGRESSION		ERROR	
x_i	y_i	$\widehat{y}_i$	$(\widehat{y}_i - \overline{y})$	$(\widehat{y}_i - \overline{y})^2$	$(y_i - \widehat{y}_i)$	$(y_i - \widehat{y}_i)^2$
60	84	100	-40	1600	-16	256
62	95	110	-30	900	-15	225
64	140	120	-20	400	20	400
66	155	130	-10	100	25	625
68	119	140	0	0	-21	441
70	175	150	10	100	25	625
72	145	160	20	400	-15	225
74	197	170	30	900	27	729
76	150	180	40	1600	-30	900

$\overline{x} = 68$ $\quad$ $\overline{y} = 140$ $\qquad\qquad$ SSR = 6000 $\qquad\qquad$ SSE = 4426

SSR MEASURES THE TOTAL
VARIABILITY DUE TO THE
REGRESSION, I.E., THE
PREDICTED VALUES OF y.
SSE WE'VE ALREADY MET.
NOTE THAT

$$\frac{SSE}{SS_{yy}}$$

IS THE PROPORTION OF
ERROR, RELATIVE TO
THE TOTAL SPREAD.

A NUMERICAL EXPRESSION FOR THE "SLOP"!

OH, GOOD!

The squared correlation

IS THE PROPORTION OF THE TOTAL SS_{yy}
ACCOUNTED FOR BY THE REGRESSION:

$$R^2 = \frac{SSR}{SS_{yy}} = 1 - \frac{SSE}{SS_{yy}}$$

(BECAUSE $SSR = SS_{yy} - SSE$). R^2 IS
ALWAYS LESS THAN 1. THE CLOSER IT
IS TO 1, THE TIGHTER THE FIT OF
THE CURVE. $R^2 = 1$ CORRESPONDS
TO PERFECT FIT.

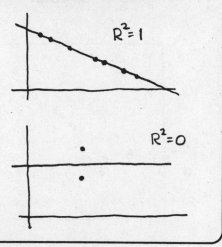

$R^2 = 1$

$R^2 = 0$

CALCULATING R^2 FOR THE
RIGGED DATA SET, WE GET

$$R^2 = \frac{6000}{10,426} = .58$$

58% OF THE VARIATION IN
WEIGHT IS EXPLAINED BY
HEIGHT. THE OTHER 42%
IS "ERROR."

PARTLY DUE TO THAT BURRITO YOU ATE!

ALTERNATELY, THE

correlation coefficient

IS THE SQUARE ROOT OF R^2 WITH THE SIGN OF b.

$$r = (\text{SIGN OF } b) \sqrt{R^2}$$

THUS, r IS + IF THE LINE GOES UP TO THE RIGHT AND − IF IT GOES DOWN TO THE RIGHT.

NEGATIVE r MEANS THAT x IS *NEGATIVELY* RELATED TO y!

r MEASURES THE TIGHTNESS OF FIT, AS WELL AS SAYING WHETHER INCREASING x MAKES y GO UP OR DOWN.

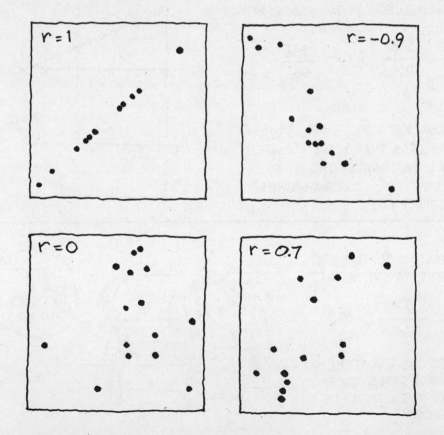

r = 1

r = −0.9

r = 0

r = 0.7

NOW LET'S BE HONEST: NOBODY— WELL, *ALMOST* NOBODY—DOES THESE CALCULATIONS BY HAND ANYMORE. WITH A COMPUTER, ALL THIS WORK CAN BE DONE IN *ONE LINE OF CODE...*

IN FACT, THIS ENTIRE BOOK CAN BE COMPRESSED INTO THE HEAD OF A STATISTICIAN...

USING THE *MINITAB* STATISTICAL SOFTWARE SYSTEM, DEVELOPED AT PENN STATE, THE SINGLE COMMAND LOOKS LIKE THIS:

```
MTB > regress 'weight' on 1 independent variable 'height'
```

AND THE RESULTS ARE

The regression equation is

WEIGHT = - 200 + 5.00 height

Predictor	Coef	Stdev	t-ratio	p
Constant	-200.0	110.7	-1.81	0.114
height	5.000	1.623	3.08	0.018

s = 25.15   R-sq = 57.5% R-sq(adj) = 51.5%

Analysis of Variance

SOURCE	DF	SS	MS	F	p
Regression	1	6000.0	6000.0	9.49	0.018
Error	7	4426.0	632.3		
Total	8	10426.0			

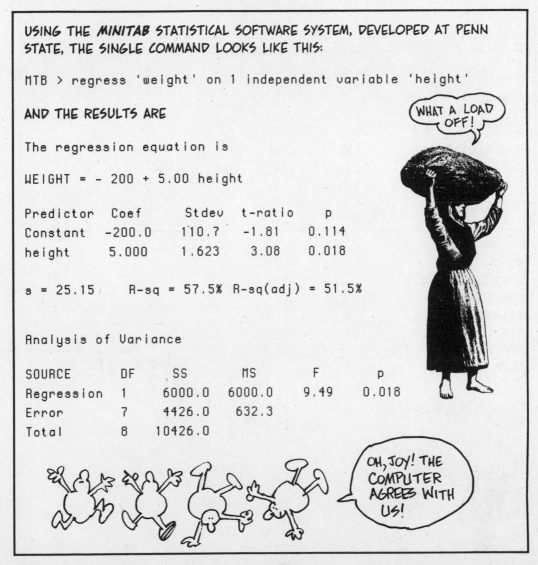

WHAT A LOAD OFF!

OH, JOY! THE COMPUTER AGREES WITH US!

NOW LET'S DO IT TO THE *REAL*
DATA OF 92 STUDENTS:

```
MTB > regress 'weight' on 1 independent variable 'height'
```

AND THE RESULTS

The regression equation is
WEIGHT = - 205 + 5.09 HEIGHT

Predictor	Coef	Stdev	t-ratio	p
Constant	-204.74	29.16	-7.02	0.000
height	5.0918	0.4237	12.02	0.000

s = 14.79   R-sq = 61.6% R-sq(adj) = 61.2%

Analysis of Variance

SOURCE	DF	SS	MS	F	p
Regression	1	31592	31592	144.38	0.000
Error	90	19692	219		
Total	91	51284			

HERE IS THE
SCATTERPLOT WITH
THE FITTED
REGRESSION LINE.
THE CORRELATION
COEFFICIENT FOR THIS
DATA SET IS

$$r = +\sqrt{.616} = .78$$

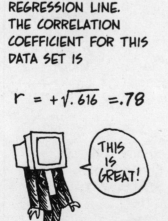

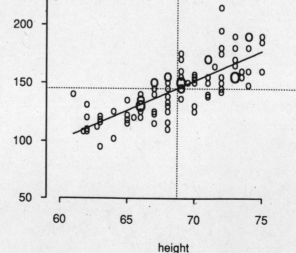

STATISTICAL INFERENCE

UP TO NOW, WE HAVE BEEN DOING *DATA ANALYSIS*, DESCRIBING THE NEAREST LINEAR RELATIONSHIP BETWEEN THE OBSERVED DATA x AND y. NOW LET'S SHIFT OUR POINT OF VIEW, AND REGARD THE 92 STUDENTS AS A *SAMPLE* OF THE POPULATION OF STUDENTS AT LARGE. WHAT CAN WE INFER?

WHAT HAPPENED TO YOU?

GOT SAMPLED.

A *REGRESSION MODEL* FOR THE WHOLE POPULATION IS A LINEAR RELATIONSHIP

$$Y = \alpha + \beta x + \epsilon$$

NOTE GREEK LETTERS TO INDICATE MODEL-DOM!

Y IS THE DEPENDENT RANDOM VARIABLE; x IS THE INDEPENDENT VARIABLE (WHICH MAY OR MAY NOT BE RANDOM); α AND β ARE THE UNKNOWN PARAMETERS WE SEEK TO ESTIMATE; AND ϵ REPRESENTS RANDOM ERROR FLUCTUATIONS.

FOR THE HEIGHT VS. WEIGHT MODEL, Y IS WEIGHT, x IS HEIGHT, α AND β ARE UNKNOWN, AND YOU CAN THINK OF ϵ AS THE *RANDOM COMPONENT* OF THE WEIGHTS Y FOR EACH VALUE OF HEIGHT x.

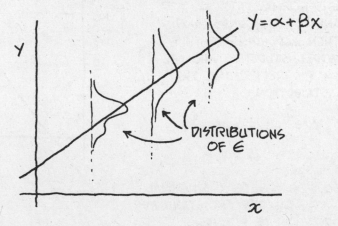

$Y = \alpha + \beta x$

DISTRIBUTIONS OF ϵ

THE DISTRIBUTION OF ϵ IS IN FACT **DIFFERENT** FOR DIFFERENT VALUES OF x: 5-FOOTERS VARY LESS IN THEIR WEIGHT THAN 6-FOOTERS. NEVERTHELESS, WE NOW MAKE A SIMPLIFYING ASSUMPTION: LET'S SUPPOSE THAT FOR ALL VALUES OF x, THE ϵ'S ARE **INDEPENDENT, NORMAL,** AND HAVE THE **SAME STANDARD DEVIATION** $\sigma = \sigma(\epsilon)$ AND MEAN $\mu = 0$.

REALITY

REALITY SIMPLIFIED

MAN! SOME OF THOSE LI'L TODDLERS MUST WEIGH LESS THAN NOTHIN'!

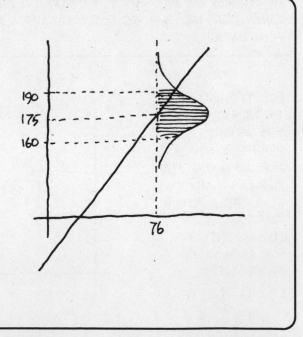

SO... MAYBE THE WEIGHT MODEL MIGHT BE

$$Y = -125 + 4x + \epsilon$$

ϵ IS NORMAL WITH $\mu = 0$ AND $\sigma = 15$ POUNDS (SAY). THEN, ACCORDING TO THIS MODEL, STUDENTS WHO ARE 6'4" (76 INCHES) HAVE THE DISTRIBUTION OF

$$Y = -125 + 4(76) + \epsilon$$
$$= 175 + \epsilon$$

SO, FOR $x = 76$, Y IS NORMAL WITH MEAN 175 AND STANDARD DEVIATION 15 POUNDS.

NOW, GIVEN THE MODEL $Y = \alpha + \beta x + \epsilon$, WE WANT TO DO AS WE'VE DONE REPEATEDLY IN THE LAST FEW CHAPTERS: TAKE A **SAMPLE** AND USE IT TO **ESTIMATE** α AND β.

ONE CAN SHOW THAT THE a AND b WE GOT BY THE LEAST-SQUARES METHOD ARE **BLUE**: THE **B**EST **L**INEAR **U**NBIASED **E**STIMATORS OF α AND β (WHATEVER THAT MEANS!).

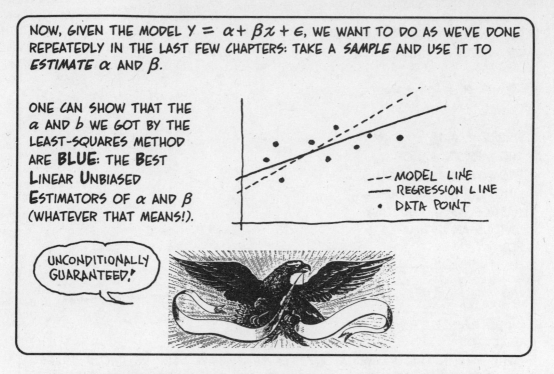

---- MODEL LINE
—— REGRESSION LINE
• DATA POINT

UNCONDITIONALLY GUARANTEED!

AS USUAL, DIFFERENT SAMPLES YIELD DIFFERENT COLLECTIONS OF DATA, WHICH GENERATE DIFFERENT REGRESSION LINES. THESE LINES ARE **DISTRIBUTED** AROUND THE LINE $Y = \alpha + \beta x + \epsilon$. OUR QUESTION BECOMES: HOW ARE a AND b DISTRIBUTED AROUND α AND β, RESPECTIVELY, AND HOW DO WE CONSTRUCT **CONFIDENCE INTERVALS** AND **TEST HYPOTHESES**?

THEY'RE **BLUE**... I'M GREEN...

LUCKILY, I ENJOY BEING GREEN...

$Y = \alpha + \beta x$

FOR EACH DATA POINT (x_i, y_i), WE HAVE

$$y_i = a + bx_i + e_i$$

WHERE $e_i = y_i - \hat{y}_i$ IS THE y-DISTANCE OF y_i FROM THE REGRESSION LINE. THE e_i ARE **SAMPLE VALUES OF** ϵ, AND THEY GIVE US AN ESTIMATOR S FOR $\sigma(\epsilon)$:

$$S = \sqrt{\frac{\sum_{i=1}^{n} e_i^2}{n-2}}$$

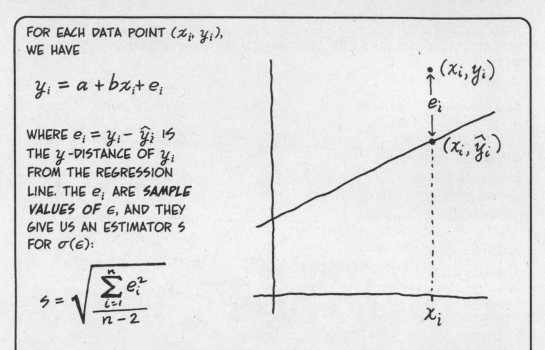

(WHY $n-2$ IN THE DENOMINATOR? BECAUSE WE HAVE USED UP TWO DEGREES OF FREEDOM TO COMPUTE a AND b, LEAVING $n-2$ INDEPENDENT PIECES OF INFORMATION TO ESTIMATE σ.)

ALTHOUGH IT ISN'T OBVIOUS, WE CAN ALSO WRITE S AS

$$S = \sqrt{\frac{SS_{yy} - b SS_{xy}}{n-2}}$$

A FORMULA WHICH ALLOWS US TO COMPUTE S DIRECTLY FROM THE SAMPLE STATISTICS.

LEARN n-DIMENSIONAL GEOMETRY, I TELL YOU, AND IT'S EASY!

TO REPEAT, S IS AN ESTIMATOR OF *HOW WIDELY THE DATA POINTS WILL BE SCATTERED AROUND THE LINE.*

confidence intervals

THE 95% CONFIDENCE INTERVALS FOR α AND β HAVE THAT OLD, FAMILIAR FORM:

$$\beta = b \pm t_{.025} SE(b)$$

$$\alpha = a \pm t_{.025} SE(a)$$

WHERE WE USE THE t DISTRIBUTION WITH $n-2$ DEGREES OF FREEDOM (FOR THE SAME REASON AS ABOVE).

THE STANDARD ERRORS, HOWEVER, LOOK RATHER UNFAMILIAR. THEY ARE (WITHOUT DERIVATION):

$$SE(b) = \frac{s}{\sqrt{SS_{xx}}}$$

$$SE(a) = s \sqrt{\frac{1}{n} + \frac{\bar{x}^2}{SS_{xx}}}$$

WHAT HAPPENED TO OUR PRECIOUS $\frac{1}{\sqrt{n}}$? IT WAS REPLACED BY SS_{xx}. LIKE n, SS_{xx} INCREASES AS WE ADD MORE DATA POINTS, BUT IT ALSO REFLECTS THE **TOTAL SPREAD OF THE x DATA**. FOR EXAMPLE, IF **ALL STUDENTS SAMPLED HAD THE SAME HEIGHT**, WE WOULD BE UNJUSTIFIED IN DRAWING ANY CONCLUSION ABOUT THE DEPENDENCE OF WEIGHT ON HEIGHT. IN THAT CASE, $SS_{xx} = 0$, GIVING $b = \infty$ AND **INFINITELY WIDE CONFIDENCE INTERVALS**.

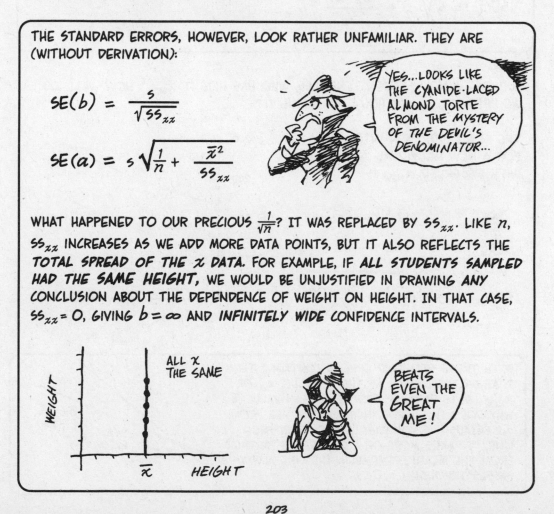

HOW WELL CAN WE PREDICT THE *MEAN RESPONSE* Y AT A FIXED VALUE x_0? FOR INSTANCE, WHAT IS THE MEAN WEIGHT OF STUDENTS OF HEIGHT 76 INCHES? THE 95% CONFIDENCE INTERVAL FOR $Y = \alpha + \beta x_0$ IS

$$\alpha + \beta x_0 = a + b x_0 \pm t_{.025} SE(\hat{y})$$

WHERE

$$SE(\hat{y}) = s\sqrt{\frac{1}{n} + \frac{(x_0 - \bar{x})^2}{SS_{xx}}}$$

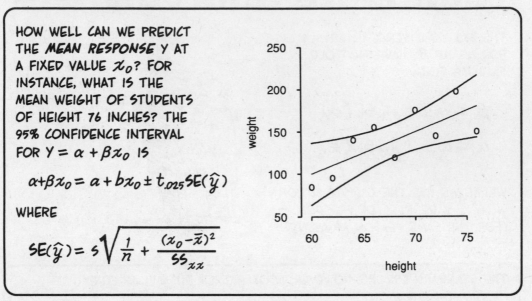

SUPPOSE A NEW STUDENT ENROLLS, WHO HAS HEIGHT x_{NEW}. HOW WELL CAN WE PREDICT Y_{NEW} WITHOUT MEASURING IT?

THE 95% PREDICTION INTERVAL FOR A NEW INDIVIDUAL Y_{NEW} WITH OBSERVED x_{NEW} IS

$$Y_{NEW} = a + b x_{NEW} \pm t_{.025} SE(Y_{NEW})$$

WHERE

$$SE(Y_{NEW}) = s\sqrt{1 + \frac{1}{n} + \frac{(x_{NEW} - \bar{x})^2}{SS_{xx}}}$$

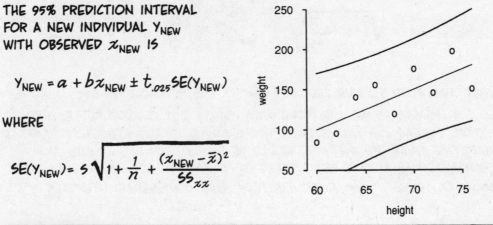

BOTH THESE STANDARD ERRORS CONTAIN A TERM THAT GROWS LARGER AS THE x-VALUE, x_0 OR x_{NEW}, GETS FARTHER FROM THE MEAN VALUE $\bar{x}$. WHY DOES THE ERROR INCREASE FARTHER FROM $\bar{x}$? BECAUSE, IF YOU WIGGLE THE REGRESSION LINE, IT MAKES MORE OF A DIFFERENCE FARTHER FROM THE MEAN! (REMEMBER, THE LINE ALWAYS PASSES THROUGH $(\bar{x}, \bar{y})$.)

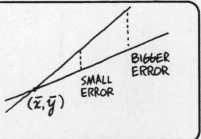

LET'S WORK IT OUT FOR THE RIGGED DATA: FOR THE **MEAN WEIGHT** WHEN $x = 76$ INCHES, WE HAVE $b = -200$ AND $a = 5$. THEN

$$Y = -200 + 5(76) \pm (2.365)(25.15)$$

$$= 180 \pm (2.365)(25.15)\sqrt{.3777}$$

$$= \mathbf{180 \pm 36.34} \text{ POUNDS}$$

THE ESTIMATED MEAN OF 6'4" STUDENTS IS 180 POUNDS, AND WE'RE 95% CONFIDENT THAT WE'RE WITHIN **36 POUNDS** OF THE TRUE MEAN.

FOR A **NEW** STUDENT WHO'S 6'4", WE USE OUR RIGGED SAMPLE OF NINE POINTS TO PREDICT THAT

$$Y_{NEW} = -200 + 5(76) \pm (2.365)(25.15)\sqrt{1 + \frac{1}{9} + \frac{(76-68)^2}{290}}$$

$$= 180 \pm (2.365)(29.51)$$

$$= \mathbf{180 \pm 70} \text{ POUNDS}$$

WE TELL THE FOOTBALL COACH THAT WE'RE PRETTY SURE THE NEW GUY WEIGHS SOMEWHERE BETWEEN **110** AND **250!!!**

THE INTERVALS ARE PRETTY TERRIBLE! WHAT'S THE PROBLEM? THERE ARE TWO PROBLEMS, ACTUALLY:

THE PENN STATE STUDENTS GIVE BETTER ESTIMATES.

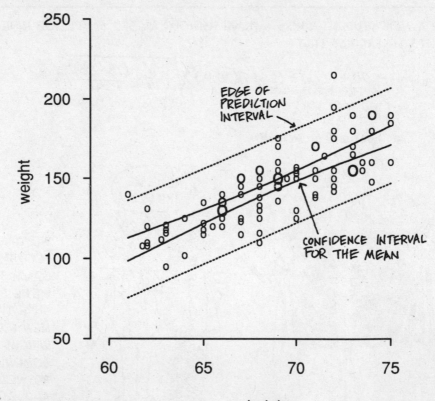

hypothesis testing

THE COMPLETE SKEPTIC MIGHT SUGGEST THAT THERE IS *NO RELATIONSHIP* BETWEEN HEIGHT AND WEIGHT. THIS AMOUNTS TO SAYING THAT $\beta = 0$.

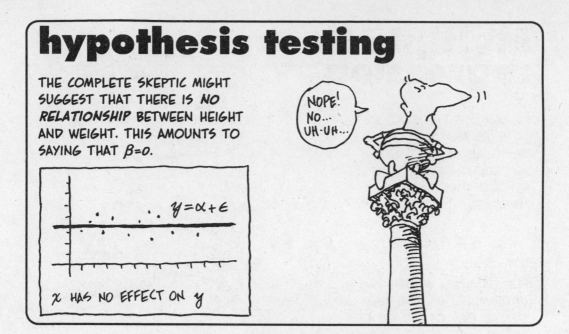

$$y = \alpha + \epsilon$$

x HAS NO EFFECT ON y

WE TAKE THIS AS THE *NULL HYPOTHESIS.*

$$H_0 : \beta = 0$$

IN THAT CASE, THE TEST STATISTIC

$$t = \frac{b}{SE(b)}$$

HAS THE t DISTRIBUTION WITH $n-2$ DEGREES OF FREEDOM. AS USUAL, THE SIGNIFICANCE TEST DEPENDS ON THE ALTERNATE HYPOTHESIS.

$$t > t_\alpha \quad \text{FOR} \quad H_a : \beta > 0$$

$$t < t_\alpha \quad \text{FOR} \quad H_a : \beta < 0$$

$$|t| > |t_{\alpha/2}| \quad \text{FOR} \quad H_a : \beta \neq 0$$

FOR THE RIGGED WEIGHT DATA, WE STRONGLY SUSPECT THE ALTERNATE HYPOTHESIS SHOULD BE

$$H_a : \beta > 0$$

WE TEST

$$t_{OBS} = \frac{5}{SE(b)} = \frac{5}{1.62}$$

$$= 3.08$$

FOR 7 DEGREES OF FREEDOM, $t_{.05} = 1.895$. SINCE $t_{OBS} > t_{.05}$, WE REJECT THE NULL HYPOTHESIS AT THE $\alpha = .05$ SIGNIFICANCE LEVEL AND CONCLUDE THAT THERE IS A SIGNIFICANT, POSITIVE RELATIONSHIP BETWEEN HEIGHT AND WEIGHT.

WHAT A SURPRISE!

Multiple linear regression

WE CAN USE THE SAME BASIC IDEAS TO ANALYZE RELATIONSHIPS BETWEEN A DEPENDENT VARIABLE AND *SEVERAL* INDEPENDENT VARIABLES:

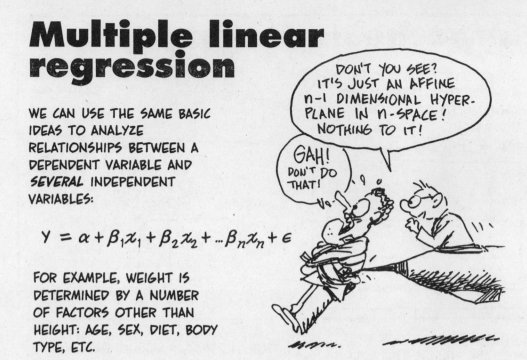

$$Y = \alpha + \beta_1 x_1 + \beta_2 x_2 + ... \beta_n x_n + \epsilon$$

FOR EXAMPLE, WEIGHT IS DETERMINED BY A NUMBER OF FACTORS OTHER THAN HEIGHT: AGE, SEX, DIET, BODY TYPE, ETC.

MATRIX ALGEBRA AND A COMPUTER COMBINE TO MAKE SUCH PROBLEMS EASY TO ANALYZE.

Non-linear regression

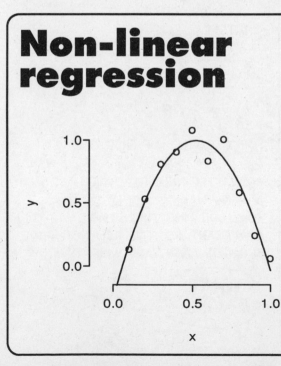

SOMETIMES DATA OBVIOUSLY FIT A *NON-LINEAR* CURVE. STATISTICIANS HAVE A BAG OF TRICKS FOR USING *LINEAR* REGRESSION TECHNIQUES FOR NON-LINEAR PROBLEMS. THE SIMPLEST OF THESE IS TO WRITE Y AS A *POLYNOMIAL*

$$Y = \alpha + \beta_1 x + \beta_2 x^2 + \epsilon$$

AND TREAT x AND x^2 AS INDEPENDENT VARIABLES IN A LINEAR MODEL.

Regression diagnostics

FITTING A COMPLEX MODEL TO DATA CAN SOMETIMES OBSCURE MANY DIFFICULTIES. WE USE REGRESSION DIAGNOSTIC PROCEDURES TO UNCOVER ANY LURKING NASTY SURPRISES.

THE SIMPLEST PROCEDURE IS TO PLOT THE *RESIDUALS* e_i AGAINST THE *PREDICTOR* y_i. REMEMBER, THE ERROR ϵ IS ASSUMED TO BE INDEPENDENT OF x.

A *RANDOM* SCATTERPLOT INDICATES THAT THE MODEL ASSUMPTIONS ARE PROBABLY OK.

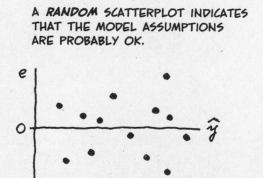

ANY *PATTERN* INDICATES A DEFINITE PROBLEM WITH THE MODEL ASSUMPTIONS.

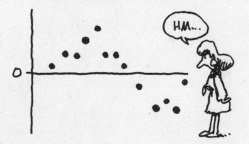

A TYPICAL LURKING NASTY SURPRISE (WHICH EXISTS IN THE HEIGHT/WEIGHT DATA) IS THAT ERRORS ARE *HETEROSCEDASTIC*: I.E., THE SPREAD OF e INCREASES AS y INCREASES.

IN THIS CHAPTER, WE HAVE SUMMARIZED THE BASIC IDEAS AND TECHNIQUES OF REGRESSION ANALYSIS, THE STUDY OF STATISTICAL RELATIONSHIPS BETWEEN VARIABLES. THIS CONCLUDES OUR DETAILED DISCUSSION OF BASIC STATISTICAL METHODS. IN OUR FINAL CHAPTER, WE'LL BRIEFLY REVIEW A FEW REMAINING TOPICS AND ISSUES.

♦Chapter 12♦
CONCLUSION

THE BASIC PRINCIPLES, TOOLS, AND CALCULATIONS COVERED IN THIS BOOK CAN BE EXTENDED TO SOLVE MORE COMPLEX PROBLEMS. HERE'S A *BIASED SAMPLE* OF MORE ADVANCED STATISTICAL METHODS!

DATA DISPLAY

WE SAW HOW TO DISPLAY *ONE* VARIABLE WITH A DOT PLOT AND *TWO* VARIABLES USING A SCATTERPLOT—BUT HOW DO WE GRAPHICALLY DISPLAY *MORE THAN TWO* VARIABLES ON A FLAT PAGE? AMONG THE MANY POSSIBILITIES, A CARTOON GUIDE HAS TO MENTION *HERMAN CHERNOFF'S* SIMPLE IDEA: USING THE HUMAN FACE, ASSIGN EACH FEATURE TO A VARIABLE AND DRAW THE RESULTING *CHERNOFF FACES*:

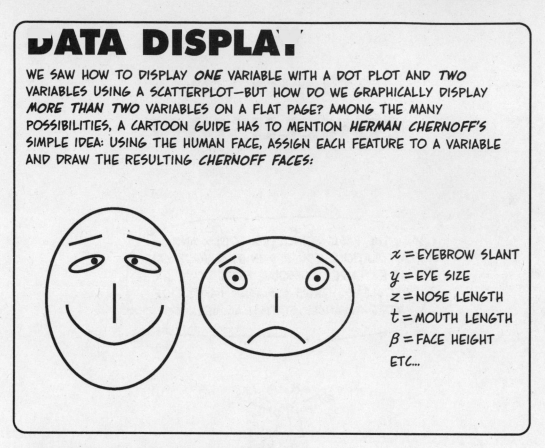

x = EYEBROW SLANT
y = EYE SIZE
z = NOSE LENGTH
t = MOUTH LENGTH
β = FACE HEIGHT
ETC...

Statistical analysis of
MULTIVARIATE DATA

AN ASSORTMENT OF MULTIVARIATE MODELS HELP TO ANALYZE AND DISPLAY n-DIMENSIONAL DATA. SOME MULTIVARIATE TECHNIQUES:

Cluster analysis

SEEKS TO DIVIDE THE POPULATION INTO HOMOGENEOUS SUBGROUPS. FOR EXAMPLE, BY ANALYZING CONGRESSIONAL VOTING PATTERNS, WE FIND THAT REPRESENTATIVES FROM THE *SOUTH* AND *WEST* FORM TWO DISTINCT CLUSTERS.

Discriminant analysis

IS THE REVERSE PROCESS. FOR EXAMPLE, A COLLEGE ADMISSIONS OFFICE MIGHT LIKE TO FIND DATA GIVING ADVANCE WARNING WHETHER AN APPLICANT WILL GO ON TO BE A *SUCCESSFUL GRADUATE* (DONATES HEAVILY TO THE ALUMNI FUND) OR AN *UNSUCCESSFUL* ONE (GOES OUT TO DO GOOD IN THE WORLD AND IS NEVER HEARD FROM AGAIN).

Factor analysis

SEEKS TO EXPLAIN HIGH-DIMENSIONAL DATA WITH A SMALLER NUMBER OF VARIABLES. A PSYCHOLOGIST MAY GIVE A TEST WITH 100 QUESTIONS, WHILE SECRETLY ASSUMING THAT THE ANSWERS DEPEND ON ONLY A FEW *FACTORS*: EXTROVERSION, AUTHORITARIANISM, ALTRUISM, ETC. THE TEST RESULTS WOULD THEN BE SUMMARIZED USING ONLY A FEW COMPOSITE SCORES IN THOSE DIMENSIONS.

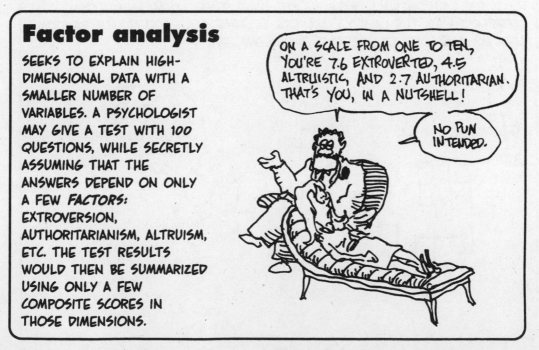

THERE IS ALSO MORE TO
PROBABILITY:

Random walks BEGIN WITH A COIN FLIP. SUPPOSE YOU MOVE AHEAD ONE STEP FOR A HEAD AND BACK ONE STEP FOR A TAIL. (USING TWO COINS, YOU CAN DO THIS IN TWO DIMENSIONS.) REPEATED FLIPS PRODUCE A STOCHASTIC PROCESS CALLED A *RANDOM WALK.* RANDOM WALK MODELS ARE USED IN *STOCK OPTION TRADING* AND *PORTFOLIO MANAGEMENT.*

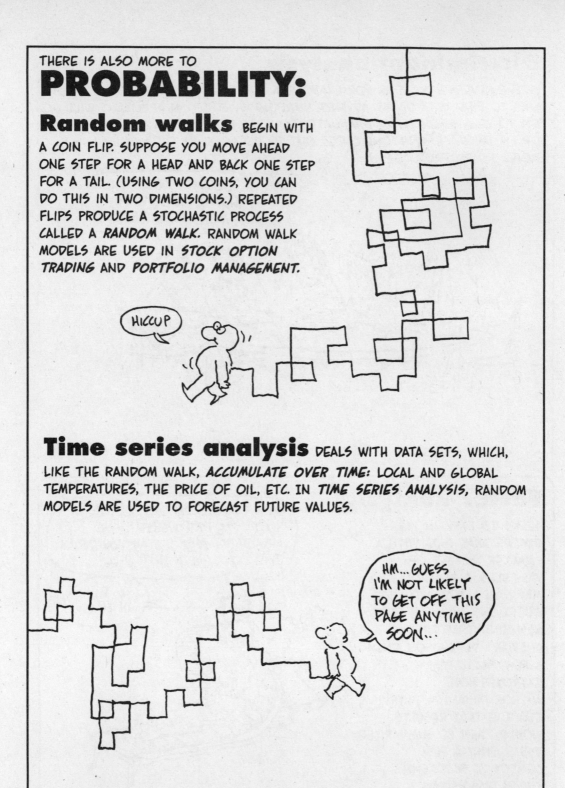

Time series analysis DEALS WITH DATA SETS, WHICH, LIKE THE RANDOM WALK, *ACCUMULATE OVER TIME:* LOCAL AND GLOBAL TEMPERATURES, THE PRICE OF OIL, ETC. IN *TIME SERIES ANALYSIS,* RANDOM MODELS ARE USED TO FORECAST FUTURE VALUES.

WE'VE ALREADY SEEN HOW THE *COMPUTER* HELPS WITH ANALYSIS AND ARITHMETIC. THERE ARE ALSO SOME STATISTICAL IDEAS THAT OWE THEIR VERY *EXISTENCE* TO THE COMPUTER:

Image analysis

A COMPUTER IMAGE MIGHT CONSIST OF 1000 BY 1000 PIXELS, WITH EACH DATA POINT REPRESENTED FROM A RANGE OF 16.7 MILLION COLORS AT ANY PIXEL. STATISTICAL IMAGE ANALYSIS SEEKS TO EXTRACT MEANING FROM "INFORMATION" LIKE THIS.

Resampling

SOMETIMES, STANDARD ERRORS AND CONFIDENCE LIMITS ARE IMPOSSIBLE TO FIND. ENTER *RESAMPLING*, A TECHNIQUE THAT TREATS THE SAMPLE *AS THOUGH IT WERE THE POPULATION*. THESE TECHNIQUES GO BY SUCH NAMES AS *RANDOMIZATION*, *JACKKNIFE*, AND *BOOTSTRAPPING*.

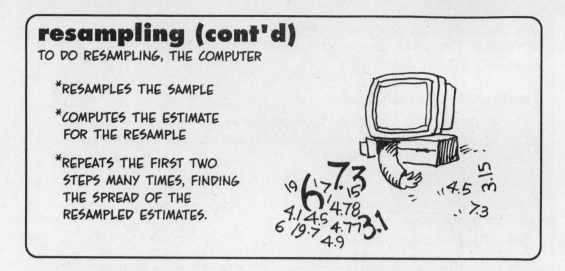

resampling (cont'd)

TO DO RESAMPLING, THE COMPUTER

- *RESAMPLES THE SAMPLE

- *COMPUTES THE ESTIMATE FOR THE RESAMPLE

- *REPEATS THE FIRST TWO STEPS MANY TIMES, FINDING THE SPREAD OF THE RESAMPLED ESTIMATES.

REMEMBER THE CORRELATION COEFFICIENT r OF THE 92 HEIGHT-WEIGHT PAIRS OF CHAPTER 11? WHAT'S THE **STANDARD ERROR OF** r? THE COMPUTER TAKES 200 BOOTSTRAP SAMPLES FROM THE 92 DATA POINTS, COMPUTES r EACH TIME, AND PLOTS A HISTOGRAM OF THE r VALUES.

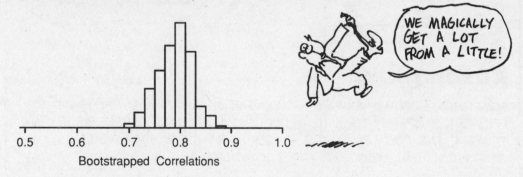

WE MAGICALLY GET A LOT FROM A LITTLE!

Bootstrapped Correlations

NOTE THAT THE SPREAD OF THE BOOTSTRAP ESTIMATES IS RELATIVELY SMALL.

AND, FINALLY, HERE ARE SOME OTHER ISSUES TO KEEP IN MIND:

DATA QUALITY

SEEMINGLY SMALL ERRORS IN SAMPLING, MEASUREMENT, AND DATA RECORDING CAN PLAY HAVOC WITH ANY ANALYSIS. *R. A. FISHER,* GENETICIST AND FOUNDER OF MODERN STATISTICS, NOT ONLY DESIGNED AND ANALYZED ANIMAL BREEDING EXPERIMENTS, HE ALSO *CLEANED THE CAGES* AND *TENDED THE ANIMALS,* BECAUSE HE KNEW THAT THE LOSS OF AN ANIMAL WOULD INFLUENCE HIS RESULTS.

MODERN STATISTICIANS, WITH THEIR COMPUTERS, DATABASES, AND GOVERNMENT GRANTS, HAVE LOST SOME OF THIS HANDS-ON ATTITUDE.

HEY, I'M GOOD TO MY MOUSE, TOO!

IF YOU GRAPHED THE MEAN MASS OF RAT DROPPINGS UNDER STATISTICIANS' FINGERNAILS OVER TIME, IT WOULD *PROBABLY* LOOK SOMETHING LIKE THIS:

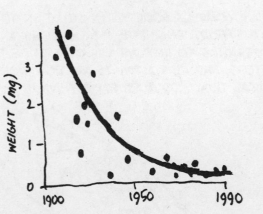

Innovation

THE BEST SOLUTIONS ARE NOT ALWAYS IN THE BOOK! FOR EXAMPLE, A COMPANY HIRED TO ESTIMATE THE COMPOSITION OF A *GARBAGE DUMP* WAS FACED WITH SOME INTERESTING PROBLEMS NOT FOUND IN YOUR STANDARD TEXT...

Communication

BRILLIANT ANALYSIS IS WORTHLESS UNLESS THE RESULTS ARE CLEARLY COMMUNICATED IN PLAIN LANGUAGE, INCLUDING THE DEGREE OF STATISTICAL UNCERTAINTY IN THE CONCLUSIONS. FOR INSTANCE, THE MEDIA NOW MORE REGULARLY REPORT THE MARGIN OF ERRORS IN THEIR POLLING RESULTS.

Teamwork

IN OUR COMPLEX SOCIETY, THE SOLUTION TO MANY PROBLEMS REQUIRES A *TEAM EFFORT*. ENGINEERS, STATISTICIANS, AND ASSEMBLY LINE WORKERS ARE COOPERATING TO IMPROVE THE QUALITY OF THEIR PRODUCTS. BIOSTATISTICIANS, DOCTORS, AND AIDS ACTIVISTS ARE NOW WORKING TOGETHER TO DESIGN CLINICAL TRIALS TO MORE RAPIDLY EVALUATE THERAPIES.

WELL, THAT'S IT! BY NOW, YOU SHOULD BE ABLE TO DO ANYTHING WITH STATISTICS, EXCEPT *LIE*, *CHEAT*, *STEAL*, AND *GAMBLE*.

WE LEFT THESE SUBJECTS TO THE BIBLIOGRAPHY!

BIBLIOGRAPHY

FOR THE STUDENT:

MOORE, DAVID S., *STATISTICS: CONCEPTS AND CONTROVERSIES*, 1991, NEW YORK, W. H. FREEMAN. EMPHASIZES IDEAS, RATHER THAN MECHANICS.

FREEDMAN, DAVID, PISANI, ROBERT, AND PURVES, ROGER, *STATISTICS*, 1991, NEW YORK, W.W. NORTON.

MOORE, DAVID S. AND McCABE, GEORGE P., *INTRODUCTION TO THE PRACTICE OF STATISTICS*, 1989, NEW YORK, W.H. FREEMAN.

SMITH, GARY, *STATISTICAL REASONING*, 1990, BOSTON, ALLYN AND BACON, INC. MORE TECHNICAL, EMPHASIZING ECONOMICS AND BUSINESS, BUT HAS EXAMPLES FROM ALL OVER.

THESE TEXTS ARE CURRENT, CORRECT, LITERATE, AND WITTY. BESIDES THE ONES WE CITE, THERE ARE HUNDREDS OF TEXTBOOKS OUT THERE, AND WE WOULD RATE MOST AS AT LEAST ACCEPTABLE.

FOR THE STRUGGLING STUDENT:

PYRCZAK, FRED, *STATISTICS WITH A SENSE OF HUMOR*, 1989, LOS ANGELES, FRED PYRCZAK PUBLISHER. AN ELEMENTARY WORKBOOK AND GUIDE TO STATISTICAL PROBLEM SOLVING

HOW TO *LIE, CHEAT, AND GAMBLE*. YOUR SAINTLY AUTHORS HAVE LITTLE EXPERIENCE IN THESE FIELDS. HERE IS SOME ADVICE FROM THE PROS:

HUFF, DARRELL, *HOW TO LIE WITH STATISTICS*, WITH PICTURES BY IRVING GEIS, NEW YORK, 1954, W.W. NORTON. CHEAP AND STILL IN PRINT!

JAFFE, A.J. AND SPIRER, HERBERT F, *MISUSED STATISTICS: STRAIGHT TALK FOR TWISTED NUMBERS*, 1987, NEW YORK, MARCEL DECKER. PART OF A GOOD POPULAR SERIES ON STATISTICS.

ORKIN, MIKE, *CAN YOU WIN?*, 1991, NEW YORK, W.H. FREEMAN. ADVICE FROM AN EXPERT ON PROBABILITY AND GAMBLING.

McGERVEY, JOHN D., *PROBABILITIES IN EVERY DAY LIFE*, 1989, N.Y., IVY BOOKS. GAMBLING FROM BLACKJACK TO SMOKING.

LAW AND SOCIETY:

GASTWIRTH, JOSEPH L., *STATISTICAL REASONING IN LAW AND POLICY, VOL. 1 & 2,* 1988, SAN DIEGO, ACADEMIC PRESS. THE LEGAL NITTY GRITTY, INCLUDING JURY SELECTION CASES LIKE THE ONE THAT BEGAN CHAPTER 9.

STEERING COMMITTEE OF THE PHYSICIANS' HEALTHY STUDY RESEARCH GROUP, *"FINAL REPORT ON THE ASPIRIN COMPONENT OF THE ONGOING PHYSICIANS' HEALTHY STUDY,"* THE NEW ENGLAND JOURNAL OF MEDICINE, VOL. 321, PP. 129-135.

IN CHAPTER 9, THE NONJUDICIAL COMMENT ON POKER FROM THE BENCH WAS FROM AN ACTUAL CASE, WE ARE ASSURED IN A PERSONAL COMMUNICATION FROM DR. JOHN De CANI, UNIVERSITY OF PENNSYLVANIA.

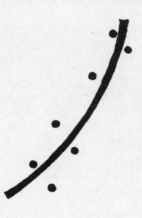

GRAPHICAL DISPLAY OF DATA:

TUFTE, EDWARD R., *THE VISUAL DISPLAY OF QUANTITATIVE INFORMATION,* 1983, CHESHIRE, CONNECTICUT, GRAPHICS PRESS.

TUFTE, EDWARD R., *ENVISIONING INFORMATION,* 1990, CHESHIRE, CONNECTICUT, GRAPHICS PRESS, THE HISTORY, ART AND SCIENCE OF GRAPHICS. BOTH BOOKS ARE CLASSICS.

CLEVELAND, WILLIAM S., *THE ELEMENTS OF GRAPHING DATA,* 1985, PACIFIC GROVE CA, WADSWORTH ADVANCED BOOKS AND SOFTWARE. DESIGN PRINCIPLES FOR COMPUTER GRAPHICS.

HISTORY:

DAVID, F. N., *GAMES, GODS AND GAMBLING,* 1962, NEW YORK, HAFNER, NEW YORK.

STIGLER, STEPHEN M., *THE HISTORY OF STATISTICS: THE MEASUREMENT OF UNCERTAINTY BEFORE 1900,* 1985, CAMBRIDGE, MA, BELKNAP PRESS OF HARVARD UNIVERSITY PRESS.

BOX, JOAN FISHER, *R. A. FISHER, THE LIFE OF A SCIENTIST,* 1978, NEW YORK, WILEY. BIOGRAPHY, BY HIS DAUGHTER, OF THE MOST INFLUENTIAL AND CONTROVERSIAL FIGURE OF 20TH CENTURY STATISTICS.

KRUSKAL, WILLIAM, *"THE SIGNIFICANCE OF FISHER: A REVIEW OF R.A. FISHER: THE LIFE OF A SCIENTIST"* 1980. JOURNAL OF THE AMERICAN STATISTICAL ASSOCIATION, VOL 75, 1030. SETS THE FISHER BIOGRAPHY IN PERSPECTIVE AND HAS EXCELLENT BIBLIOGRAPHY.

STATISTICAL SOFTWARE:

IN THIS BOOK WE USED THE *MINITAB* STATISTICAL SOFTWARE SYSTEM (MINITAB INC., STATE COLLEGE, PA). THE PENN STATE STUDENT HEIGHT AND WEIGHT DATA IS FROM THE *PULSE* DATA SET ON THIS SYSTEM. COMPUTER GRAPHICS WERE GENERATED BY *S-PLUS* (STATISTICAL SCIENCES INC, SEATTLE WA), ON A 486 PC CLONE. *S* IS SOPHISTICATED SOFTWARE, DEVELOPED BY AT&T BELL LABS FOR ADVANCED ANALYSIS AND GRAPHICAL DISPLAYS.

RYAN, BARBARA, JOINER, BRIAN, AND RYAN, THOMAS, *MINITAB HANDBOOK*, (PWS-KENT, BOSTON, 1985) AND *THE STUDENT EDITION OF MINITAB* (ADDISON WESLEY) ARE FAST, INEXPENSIVE INTRODUCTIONS TO STATISTICAL COMPUTING. MINITAB RUNS ON MAINFRAMES, PC COMPATIBLES, AND MACINTOSH COMPUTERS.

THERE ARE MANY HIGH QUALITY SOFTWARE PACKAGES AVAILABLE FOR THE PERSONAL COMPUTER, INCLUDING:

DATADESK (DATA DESCRIPTION, ITHACA, NY), FOR THE MACINTOSH

SAS (SAS INSTITUTE INC, CARY, NC), *SPSS* (SPSS INC, CHICAGO, IL), AND *BMDP* (BMDP STATISTICAL SOFTWARE, INC., LOS ANGELES, CA) WERE ORIGINALLY DESIGNED FOR MAINFRAME SYSTEMS AND NOW HAVE MIGRATED TO THE PC, COMPLETE WITH WINDOWS.

STATGRAPHICS (STATISTICAL GRAPHICS CORP, PRINCETON, NJ), FOR THE PC.

STATVIEW (ABACUS CONCEPTS, OAKLAND CA) FOR THE MACINTOSH.

SYSTAT (SYSTAT, INC., EVANSTON IL) HAS SYSTEMS THAT RUN IN ALL ENVIRONMENTS.

THESE PACKAGES DIFFER IN IMPORTANT DETAILS; YOU NEED TO BE A SMART SHOPPER. WE RECOMMEND CHOOSING A SYSTEM THAT YOUR COLLEAGUES HAVE ALREADY TESTED. FEW OF US ARE CUT OUT TO BE STATISTICAL SOFTWARE PIONEERS. WHEN LEARNING A NEW SYSTEM, EXPERIMENT WITH SMALL, FAMILIAR DATA SETS. REMEMBER, *THE MOST EXPENSIVE PART OF ANY SOFTWARE IS YOUR TIME.* THE CARTOON RULE FOR LEARNING STATISTICAL COMPUTING IS: FAMILIARITY BREEDS RESULTS.

TRYING TO LEARN STATISTICAL *THEORY* AND STATISTICAL *COMPUTING* AT THE SAME TIME IS A LITTLE LIKE TRYING TO *WALK* AND *CHEW GUM* AT THE SAME TIME. DIFFERENT SKILLS AND THOUGHT PROCESSES ARE INVOLVED IN EACH. SET ASIDE SEPARATE TIMES TO LEARN THESE SUBJECTS, THEN BRING THEM TOGETHER. IN THIS WAY, YOU CAN BECOME A *CHEWING, WALKING, COMPUTING, RENAISSANCE STATISTICIAN!*

INDEX

ABOUT THE AUTHORS

WOOLLCOTT SMITH IS PROFESSOR OF STATISTICS AT TEMPLE UNIVERSITY. WITH A B.S. AND M.A. FROM MICHIGAN STATE UNIVERSITY, AND A PH.D. FROM JOHNS HOPKINS, HE IS THE AUTHOR OR COAUTHOR OF MORE THAN FORTY PUBLICATIONS IN SUCH DIVERSE AREAS AS OIL SPILL SURVEYS, STATISTICAL THEORY, AND ENVIRONMENTAL STATISTICS. HE HAS ADVISED A NUMBER OF NATIONAL SCIENTIFIC PROGRAMS. HE DEBATES AND KAYAKS WITH HIS WIFE, LEAH, AND HIS TWO GROWN CHILDREN, KESTON AND AMELIA.

"CIVILIZATION ADVANCES BY EXTENDING THE NUMBER OF OPERATIONS WHICH WE CAN PERFORM WITHOUT THINKING ABOUT THEM."
—ALFRED NORTH WHITEHEAD

LARRY GONICK IS THE AUTHOR OR COAUTHOR OF A NUMBER OF BOOKS OF NON-FICTION CARTOONING, AS WELL AS THE FEATURE SCIENCE CLASSICS, APPEARING BIMONTHLY IN DISCOVER MAGAZINE. A DROP-OUT OF HARVARD GRADUATE SCHOOL IN MATHEMATICS, HE SEEMS TO HAVE COME FULL CIRCLE. HE LIVES IN SAN FRANCISCO WITH HIS WIFE, LISA, AND TWO DAUGHTERS, SOPHIE AND ANNA, AND HOPES TO UNDERSTAND LIFE WHILE REMAINING CHAINED TO HIS DRAWING BOARD.